Philosophy
of Science
Key Concepts

ALSO AVAILABLE FROM BLOOMSBURY

The Bloomsbury Companion to the Philosophy of Science, edited by
Steven French and Juha Saatsi
The History and Philosophy of Science: A Reader, edited by
Daniel J. McKaughan and Holly Vande Wall
Philosophy of Science: The Key Thinkers, edited by
James Robert Brown

Philosophy of Science

Key Concepts

Second edition

STEVEN FRENCH

Bloomsbury Academic
An imprint of Bloomsbury Publishing Plc

B L O O M S B U R Y
LONDON · OXFORD · NEW YORK · NEW DELHI · SYDNEY

Bloomsbury Academic

An imprint of Bloomsbury Publishing Plc

50 Bedford Square	1385 Broadway
London	New York
WC1B 3DP	NY 10018
UK	USA

www.bloomsbury.com

BLOOMSBURY and the Diana logo are trademarks of Bloomsbury Publishing Plc

First published 2016
Reprinted 2016

British Library Cataloguing-in-Publication Data
A catalogue record for this book is available from the British Library.

ISBN: HB: 978-1-4742-4524-1
PB: 978-1-4742-4523-4
ePDF: 978-1-4742-4526-5
ePub: 978-1-4742-4525-8

Library of Congress Cataloging-in-Publication Data
French, Steven.
[Science]
Philosophy of science : key concepts / Steven French.-- 2nd edition.
pages cm
Includes bibliographical references and index.
ISBN 978-1-4742-4524-1 – ISBN 978-1-4742-4523-4 1. Science–Philosophy. I. Title.
Q175.F826 2015
501–dc23
2015028292

Typeset by Fakenham Prepress Solutions, Fakenham, Norfolk, NR21 8NN
Printed and bound in Great Britain

Contents

Acknowledgements

Like the first, the second edition of this book has taken much longer than I expected or hoped to emerge blinking into the daylight. That it made it at all is testament to the support and nagging powers of a series of editors at first Continuum and then Bloomsbury, the most recent being Colleen Coalter. That I kept at it, despite all the doubts and distractions is, as always, due to the love and support and tolerance of my family, but especially Dena. And, as before, that it has the form and content that it does is due to the many Leeds first-year students I have inflicted this material on over the years. I'd like to thank all of them, together with my colleagues and the School's administrative staff, who have made it possible for me to write this. I'd especially like to give a big thank you to all the postgraduates who covered the tutorials and my former students and colleagues who also taught on the modules 'How Science Works' and 'Introduction to the Philosophy of Science' at various times: Otávio Bueno, Angelo Cei, Anjan Chakravartty, Grant Fisher, Kerry McKenzie and Juha Saatsi. I owe you all a huge debt (and no, I'm still not going to pay it off with a share of the royalties!).

1

Introduction

It is always good to begin a book like this with a statement that surely everyone will agree with: as a cultural phenomenon, science has had more of an impact on our lives than any other. We could just list the technological spin-offs alone: genetic engineering, nuclear weapons, a cure for ovarian cancer, the laptop I'm writing this on, the microwave oven I cooked my dinner in, the iPhone I take photos with and listen to my (unfashionable) music on … And of course the way in which such technologies are spun off from science is an interesting issue in itself, one which we do not have space here to tackle. But over and beyond the practical benefits, there is the profound way in which science has shaped and changed our view of the world and our view of our place in it: think of the theory of evolution and the way it has changed our understanding of our origins. Consider the further, related development of the theory of genetics: and how that has transformed not only our understanding of a range of diseases and disorders, but also our view of our behaviour, our attitudes and of ourselves. Or think of quantum physics and the claim that reality is somehow fundamentally random; or Albert Einstein's theory of relativity, according to which time runs slower the faster we move, and space and time are replaced by space-time, which is curved and distorted by the presence of matter.

Science is an amazing phenomenon, and has had a huge impact on human society over hundreds of years – so how does it work? How do scientists do the things they do? How do they come up with the theories? How do they test them? How do they use these theories to explain phenomena? How do they draw conclusions from them about how the world might be? These are the sorts of questions we'll be looking at here.

How should we go about answering them? How should we go about discovering how science works?

One approach would be to pay attention to what scientists themselves say about their work: that is, to listen to the scientists. The problem with that is that scientists often have very different and sometimes downright contradictory views about how science works. Consider, for example, an apparently quite plausible statement:

> Science is a structure built on facts.
> (J. J. Davies, *On the Scientific Method*, Harlow: Longman,
> 1968, p. 8)

Indeed, this is perhaps how many of us would begin to characterize science. It is surely what makes it distinctive and different from certain other human activities such as art, say, or poetry, or more controversially perhaps, religion. But now consider this admonition from Ivan Pavlov, famous for his experiments with the salivating dogs (which demonstrated how certain forms of behaviour can be triggered by appropriate stimuli):

> Do not become archivists of facts. Try to penetrate to the secret of their occurrence, persistently search for the laws which govern them.
> (I. Pavlov, 'To the Academic Youth of Russia', 27 February 1936,
> repr. in G. Seldes, *The Great Thoughts*, New York: Ballantine
> Books, 1996, pp. 360–1)

Now this might not seem to be in direct conflict with the previous statement; after all, Pavlov is simply urging us not to become obsessed with collecting facts, but to search for the laws underpinning them, and that can be taken to be quite consistent with the claim that science is built upon these facts (we might see facts as sitting at the base of a kind of conceptual pyramid with theoretical laws, perhaps, sitting at the top). William Lawrence Bragg, who did some fundamental work with the use of x-rays to reveal the structure of materials (some of it performed near my place of work at the University of Leeds), went a bit further by insisting that,

The important thing in science is not so much to obtain new facts as to discover new ways of thinking about them.
(W. L. Bragg, in H. J. Eysenck, *Extracted from Genius: The Natural History of Creativity*, Cambridge: Cambridge University Press, 1995, p. 1)

This kind of view meshes nicely with the view that scientific 'facts' are rock solid in some sense, that they underpin the much vaunted objectivity of science. But then here is the late Stephen Jay Gould, a much-revered professor of geology and zoology, defender of the theory of evolution and commentator on science:

In science, 'fact' can only mean 'confirmed to such a degree that it would be perverse to withhold provisional assent'. I suppose that apples may start to rise tomorrow, but the possibility does not merit equal time in physics classrooms.
(S. J. Gould, 'Evolution as Fact and Theory', in S. J. Gould, *Hen's Teeth and Horse's Toes: Further Reflections in Natural History*, 1994 [Originally published 1983], New York: W. W. Norton, p. 255)

This suggests that the 'facts' are not to be taken as the bedrock of the structure of science. On Gould's view they are the sort of things about which we might give or withhold assent and, in that giving and taking away, their status may change: yesterday's 'fact' might become today's misunderstanding, misinterpretation or downright mistake. We shall return to this issue in Chapters 4, 5, 6 and 7.

More radically, perhaps, Einstein maintained that 'If the facts don't fit the theory, change the facts.' What he meant here, is that in some cases our belief that a given theory is correct, or true in some sense, is so strong that if the 'facts' don't fit, we should conclude there is something wrong with them, rather than with the theory. And clearly there are examples from the history of science of theories that are so well entrenched that the first (and second and third ...) reaction to an apparent anomalous experimental fact would be to question the fact (or the experimenter who produced it!). Some scientists and philosophers of science would abhor such an attitude, arguing that allowing theories to become so entrenched would be to sound the death knell of science itself.

That might seem a bit melodramatic, but we can surely understand the concern: how can science progress if certain theories become so well established that they are viewed as pretty well inviolable? I don't actually think this happens in practice; rather facts that don't fit with such theories are subjected to extra-critical scrutiny, but if they survive that, then the theory itself may come to be seen as flawed. Nevertheless, it is not as straightforward as Einstein, again, seemed to think, asserting elsewhere that,

No amount of experiments can ever prove me right; a single experiment may at any time prove me wrong.
(A. Einstein, in A. Calaprice, *The New Quotable Einstein*, Princeton: Princeton University Press and Hebrew University of Jerusalem, 2005, p. 291)

This is a view – known as 'falsificationism', which holds that the crucial role of facts is not to support theories but to refute and falsify them, since in that way science may progress – to which we shall return, again, in later chapters; for the moment, let's just note how Einstein appears to have contradicted himself! Another great physicist expressed what he saw as the interplay between theory and experiment as follows:

The game I play is a very interesting one. It's imagination in a straightjacket, which is this: that it has to agree with the known laws of physics. ... It requires imagination to think of what's possible, and then it requires an analysis back, checking to see whether it fits, whether it's allowed, according to what's known, okay?
(R. Feynman, quoted in L. M. Krause, *Quantum Man: Richard Feynman's Life in Science*, New York: W. W. Norton, 2012, p. 263)

Returning to our question of how science works, a better way, I would suggest, of getting to grips with it is not to pay too much attention to what scientists themselves say, but to look at scientific practice itself. Of course, this is complex and multi-faceted and just plain messy but rather than considering how scientists *think* science works, we should look at what they *do*. This raises the further question of how we should do that.

Some philosophers and sociologists of science have suggested that if we want to know how science works we should actually go into a lab or a theoretician's office and observe how science is actually practised. This is an interesting suggestion and some sociologists have indeed approached the observation of experimental scientists in the laboratory as if they were anthropologists observing the rituals and behaviour of some tribe with a culture very different from our own. Typically, such sociologists have insisted that they went in without presuppositions, or rather, that they recorded their observations as if they had no presuppositions about the work being carried out in the lab.

But of course that is nonsense; presuppositions cannot just be left at the door and even anthropologists do not do that. Furthermore, the procedure we adopt in examining scientific practice might depend on the questions we want to ask. As we shall see, our basic question asked here – How does science work? – will be broken down into a further series: How are theories discovered? How are they supported, or not, by the evidence? What do they tell us about the world, if anything? What are the roles played by social and political factors in scientific practice? Except for the last, it is not clear how simply observing scientists in their natural habitats is going to cast much light on these issues.

And finally, most of us don't have either the inclination or the time to pursue such an approach (if you're interested in how a similar exercise might be carried out by a philosopher, consider the account by a well-known philosopher of science of his time spent in a high energy physics lab, in Ronald N. Giere's book *Explaining Science* (Chicago: University of Chicago Press, 1988); you might like to ask yourself to what extent this actually illuminates scientific practice). Instead, we look at case studies, some drawn from the history of science, some drawn from our own examination of the notebooks, records and papers of practising scientists. On the basis of such an examination, we can describe at least a certain aspect of scientific practice and, with that in hand, might start to formulate an answer to the above questions.

Now, I don't have the space to go into a huge amount of detail on these case studies here but I will draw on certain well-known (and perhaps not so well-known) episodes from current and past scientific practice to illustrate the points I want to make. Of course, you might feel that my descriptions of these episodes are too crude, too fragmentary or just too unclear to offer much in the way of illumination (and I'm sure my colleagues in the history of science will feel that way); that's fine, and I hope if you feel that's the case then you will be encouraged to examine these case studies yourself, or even come up with some of your own. The claims I make in this book are by no means definitive: there is much more to do and develop and I hope the readers and students who use this book will add to these further developments.

There is one final point before we move on to the issues themselves, and that is that some might insist that the really important question is not how does science work, but how *should* it work? In other words, what philosophers of science and commentators in general should be concerned with is not merely describing what scientists do, how they come up with their theories and test them etc., but actually specifying what they should be doing, by setting down certain *norms* of what counts as good science for example.

For many years, particularly in the first half of the twentieth century, this was taken to be an acceptable goal for the philosophy of science. Many philosophers and commentators on science saw themselves as in the business of spelling out what counted as good science, of delimiting it from bad or fake science and of effectively telling scientists what they should do in order to produce good science. Now you might say, straightaway, 'What gives them the right?!' On what grounds can philosophers and others (but especially *philosophers*!) tell scientists how they should do their work? We can take the sting out of such questions and expressions of outrage(!) by recalling that for many hundreds of years science was not regarded as distinct from philosophy, that it was indeed called 'natural philosophy' and that it was only in the late nineteenth and early twentieth centuries that the huge cultural impact of science, through technology and otherwise, and its transformative potential began to be made apparent. It's a bit of a crude overstatement but it's not too far from the truth to say that it was only with the demonstration of science's capabilities for warfare, for the development of new weapons, new defences and so on, that governments and politicians in general began to take it seriously and as worthy of significant funding.

Setting aside the technological and material impact of science and just considering the conceptual transformations it has promoted, the changes to our worldview, even here science was not particularly regarded as something special or authoritative. We can go back and look at the great debates in the nineteenth century following the publication of Charles Darwin's *On The Origin of Species* – debates which still echo down through the years – to see how science, or at least this aspect of it, came under attack. Or, we can take an 'iconic' event in the history of twentieth-century science, one we shall return to in later chapters – the British astronomer Arthur Eddington's observation of the 'bending' of starlight around the sun which confirmed Einstein's claim that space-time could be curved and distorted by massive bodies (like stars). For reasons I shall touch on later, this apparent confirmation of a technically difficult and conceptually challenging theory in physics became *the* hot news of the day, making the headlines of the major newspapers and elevating Einstein in status from an obscure Swiss-German physicist to the crazy-haired representative of science in general. Yet Einstein's theories were rejected, often with derision, by many commentators (even scientists themselves were cautious and it is worth noting that he didn't get the Nobel Prize for his relativity theory but for his early work on an aspect of quantum physics). Indeed, a famous group of philosophers got together in the 1920s and published a tract declaiming Einstein's theories as clearly false since our conceptions of space and time were bound up with the very mental framework by which we came to understand and make sense of the world and in that framework space and time simply could not be 'curved'. Einstein himself was less bothered by such claims (he famously responded with the remark, consistent with the falsificationist attitude noted above, that 'If I were wrong, one would be enough') than the anti-Semitic attacks of certain Nazi sympathizers but they illustrate how even what we now take to be major scientific advances were resisted and even rejected.

It is in this context that certain philosophers of science took on the role of defending science, of pointing out what they considered to be good science, of using that to demarcate science from what they called 'pseudo-science' (we'll come back to this in later chapters, but astronomy would count as science and astrology as pseudo-science) and of laying out what they considered to be the norms of good scientific practice. On what were these norms based? Well, in part on what these philosophers of science took to be – in modern day management-speak – 'best practice'; so, Einstein's theory and Eddington's apparent confirmation of it typically feature in these accounts as exemplars of such practice, as we'll see later. But in part the norms of good science were shaped by certain broad values to do with objectivity and rationality in general, themselves tied up with the testability of scientific theories.

However, it was the problems associated with defending these notions of objectivity and testability that led philosophers to drop out of the game of explicating how science *should* work and to concentrate on describing how it actually does. According to recent commentators, that has left a huge gap in the non-scientific public's ability to exercise some control of the agenda of science, leaving the field open to governments, multinationals and the like. Here's one such commentator who laments the loss of a normative element in these discussions:

> ... scientists must acquire a competence in the consummate democratic art of negotiation – especially with a public who will bear the financial costs and sustain the eventual impacts of whatever research is commissioned. But perhaps more important, scientists must realize that the value dimensions of their activities extend not only to the capacity of their research to do good or harm but also to the opportunity costs that are incurred by deciding to fund one sort of research over another – or, for that matter, over a non-scientific yet worthy public works project. In short, part of the social responsibility of science is to welcome the public's participation in setting the priorities of the research agenda itself.
> (S. Fuller, 'Can Science Studies Be Spoken in a Civil Tongue?',
> *Social Studies of Science* 24 (1994): 143–68)

I'm not going to get into the details of that debate here. Instead, all I'm going to do is to try to illuminate certain aspects of scientific practice in the hope that this may lead to a better appreciation of how science does, actually, work. And if anyone reading this finds it useful in helping to think through the issues involved in determining how science should work, then that's all to the good.

2

Discovery

When people think of scientists, they usually think of a man (typically) in a white coat; and when they think of what scientists do, they generally think of them making some great discovery, something for which they might be awarded the Nobel Prize. Discovery – of some fact, of some explanation of a phenomenon, of, again typically, some theory or hypothesis – is seen as lying at the heart of scientific practice. So, the fundamental question we will try to answer in this chapter is, how are scientific theories, hypotheses, models etc. *discovered*? Let's begin with a very common and well-known answer.

Common view: The 'Eureka moment'

In cartoons, creativity is often signified by a lightbulb going on over the head of the hero. It is supposed to represent the flash of inspiration. Scientific discoveries are likewise typically characterized as occurring suddenly in a dramatic creative leap of imagination, a flash of insight or a kind of 'aha!' experience. The classic example is that of Archimedes, the great Greek scientist of the third century BC, who, famously, was asked by the King of Syracuse to determine if a wreath he'd been given as a present was real gold or, somehow, fake. (The King wished to consecrate the wreath to the gods and of course it wouldn't do if it were anything other than pure gold. And because it was to be consecrated, it couldn't be opened up or analysed.) The wreath seemed to weigh the same as one made of solid gold, but that, of course, wasn't enough. Archimedes is supposed to have been visiting the public baths when he noticed that as he relaxed into the bath, the water overflowed and the deeper he sank, the more water flowed out. He realized that the water displaced could be used to measure the volume of the object immersed, and if the wreath were pure, that volume would be equal to that of an equal weight of pure gold; if not, if it were adulterated with an equal weight of, say, silver or lead, which has a different density from gold, then the volume would be different. At that point, Archimedes is reputed to have leapt from the bath and run naked through the streets, shouting 'Eureka!' or 'I've found it!' (As it turned out, the volume was greater than the same weight of pure gold and the King realized he had been cheated.)

This might seem an old, outdated story. But here's Professor Lesley Rogers, a world famous neuro-biologist:

> A visitor to my lab, doing some labelling of neural pathways with these tracer dyes, happened to think, 'Well, let's give it a go.' And when we saw it, that was a Eureka moment. Yet it was chance – he happened to come, he was looking at something entirely different, I offered him the place in the lab, we then decided to just give it a go, and it turned out.
>
> (Interview with Prof. Lesley Rogers, Australian Academy of Science, 2001; available at https://www.science.org.au/node/328009)

Here's another example from the history books: observing an apple falling from a tree, the great mathematician and scientist Isaac Newton (supposedly) realized that the force that accelerates the apple towards the ground (namely the force of gravity) is the same as the force that keeps the moon in its orbit around the earth, which is the same as the force that keeps the earth around the sun and so on. And so, it is suggested, Newton discovered his famous Law of Universal Gravitation via the 'Eureka moment' of seeing an apple fall.

Another notable and more recent example is that of Kary Mullis, who won the Nobel Prize in 1993 for his discovery of the 'polymerase chain reaction'. This is a technique that allows you to identify a strand of DNA that you might be interested in and make vast numbers of copies of it comparatively easily (and by vast, I mean *vast* – from one molecule, PCR can make 100 billion copies in a few hours). It is this which lies behind genetic 'fingerprinting', made famous through the CSI TV series for example, and it has become a standard technique in molecular biology, leading to a huge number of other applications and research results. Here is Mullis's own recollection of the discovery, made, he claims, as he drove up through the hills of northern California, with the smell of buckeye blossom in the air and a new idea in his mind:

My little silver Honda's front tires pulled us through the mountains. My hands felt the road and the turns. My mind drifted back into the lab. DNA chains coiled and floated. Lurid blue and pink images of electric molecules injected themselves somewhere between the mountain road and my eyes.

I see the lights on the trees, but most of me is watching something else unfolding. I'm engaging in my favourite pastime.

Tonight I am cooking. The enzymes and chemicals I have at Cetus [his lab] are my ingredients. I am a big kid with a new car and a full tank of gas. I have shoes that fit. I have a woman sleeping next to me and an exciting problem, a big one that is in the wind.

"What cleverness can I devise tonight to read the sequence of the King of molecules?"

DNA. The big one. ...

[He then describes how he thought of the problem in terms of a 'reiterative mathematical procedure' which would allow him to find a specific sequence of DNA and he then realized that he could use a short piece of DNA itself to do this and then initiate a process of reproduction, using the natural properties of DNA to replicate itself. Then the lightbulb went off ...]

'Holy shit!' I hissed and let off the accelerator. The car coasted into a downhill turn. I pulled off. ... We were at mile marker 46.58 on Highway 128 and we were at the very edge of the dawn of the age of PCR. ... I would be famous. I would get the Nobel Prize. [And he did.]

(K. Mullis, *Dancing in the Mind Field*, London: Bloomsbury, 1999, pp. 3–7)

[We might be a little suspicious that he retained enough presence of mind on this momentous occasion to actually note where he was along the highway. A short video clip of Mullis describing the discovery can be found at www.dnai.org/text/204_making_many_dna_copies_ kary_mullis.html. Another Nobel Prize winner, the physicist Steven Weinberg, also made the central discovery that won him the prize while driving to MIT in his red Camaro. These are the only examples I know of scientific discoveries made while driving a car!]

This is a compelling view of scientific discovery. It chimes with widely held accounts of creativity in general which hold that it all comes down to similar 'Eureka' or, less classically, 'Holy shit!' moments. In particular it is consistent with a similar view of art, which takes the artist to be in the grip of her 'muse' who strikes her with (divine) inspiration. Another famous example is that of Mozart, who in the play and movie *Amadeus* is portrayed as this wild and foul-mouthed creative genius, capable of composing a brilliant piece of music on the spot and the subject of murderous envy by plodding Salieri, who has devoted years and years to studying his craft, yet clearly has less talent in his whole body than Mozart has in his little finger! This is a beguiling picture and perhaps its attractiveness helps to explain why scientists themselves are so keen to present their discoveries as fitting into this 'Eureka' mould.

This view of discovery also fits with a historically well-entrenched approach to creativity in general, known as the 'Romantic' view.

The 'Romantic' view of creativity

Here is a classic statement of the 'Romantic' view by the great German philosopher, Immanuel Kant:

We thus see (1) that genius is a *talent* for producing that for which no definite rule can be given; it is not a mere aptitude for what can be learnt by a rule. Hence *originality* must be its first property. (2) But since it also can produce original nonsense, its products must be models, *i.e. exemplary*; and they consequently ought not to spring from imitation, but must serve as a standard or rule of judgment for others. (3) It cannot describe or indicate scientifically how it brings about its products, but it gives the rule just as nature does. Hence the author of a product for which he is indebted to his genius does not know himself how he has come by his Ideas; and he has not the power to devise the like at pleasure or in accordance with a plan, and to communicate it to others in precepts that will enable them to produce similar products.

(I. Kant, *The Critique of Judgment*, 1790)

There are a couple of things to note about this description. First of all, it states that creativity involves 'no definite rule'; that is, it cannot be analysed or described in terms of some method. Second, as a consequence even the discoverer doesn't know how she made the discovery. In other words, discovery is ultimately *irrational* and *unanalysable*.

This 'common view' of discovery is then used to underpin an important distinction, one that apparently helps us to understand scientific practice. This is the distinction between the context in which discovery takes place, and the context in which justification, or the impact of evidence, takes place.

Context of discovery vs. context of justification

The idea here is to separate out those aspects of scientific practice that are irrational and creative from those that are rational and, possibly, rule governed. The former falls under what is known as the 'context of discovery', the latter under the 'context of justification'. Let's look at this distinction in detail.

Context of discovery

Because discovery is 'creative' and irrational, it is not open to investigation by philosophers who are interested in what is rational about science. As we saw above, according to the 'Romantic' view it involves no definite rule, but it does involve talent or genius. Perhaps it is going too far to say that it is 'unanalysable', as psychologists have written reams on creativity and the origins of genius. Furthermore, there is evidence that particularly creative moments occur under certain conditions: of calmness and relaxation, for example (think of Archimedes in his bath or Mullis driving along the highway). Such moments might be open to investigation by sociologists (or, in Mullis's case, drug counsellors!). So, the context of discovery covers those aspects of scientific practice when discovery takes place – the Eureka moments, the creative leaps, the flashes of insight. It is not philosophically analysable, because philosophy is concerned with what is rational, but it is analysable by psychologists and sociologists.

Context of justification

This is concerned with the rational features of scientific practice, and particularly with the issue of how theories are *justified*, or supported, by the evidence. This *is* open to investigation by philosophers because it covers what is rational about science. We shall look at justification and the role of evidence in science in Chapter 5 but all we wish to emphasize here is the difference between this and the context of discovery. Here is how one of the most famous philosophers of science of the twentieth century put it:

> ... the work of the scientist consists in putting forward and testing theories.
>
> The initial stage, the act of conceiving or inventing a theory, seems to me neither to call for logical analysis nor to be susceptible of it. The question how it happens that a new idea occurs to a man – whether it is a musical theme, a dramatic conflict, or a scientific theory – may be of great interest to empirical psychology; but it is irrelevant to the logical analysis of scientific knowledge ... My view of the matter ... is that there is no such thing as a logical method of having new ideas, or a logical reconstruction of this process. My view may be expressed by saying that every discovery contains 'an irrational element', or a 'creative intuition'...
>
> (K. Popper, *The Logic of Scientific Discovery*, New York: Basic Books, pp. 31–2).

This seems a pretty intuitive distinction and it in turn meshes with a more general account of scientific method known as the 'hypothetico-deductive' account, or the 'method of hypothesis'. Let's now look at this.

The hypothetico-deductive account

This gets its name in the following way:

Hypothetico This indicates that hypotheses are generated through creative leaps, Eureka moments, drug-addled visions, whatever.

Deductive Experimental consequences are then deduced from the hypothesis and are subjected to experimental testing. By 'deduced', here we mean by the rules of logical deduction, as spelled out in all good logic textbooks. If these implications turn out to be correct the hypothesis is said to be confirmed; if not, it is falsified. As I indicated, we'll look at this aspect of scientific practice more closely in Chapter 5 but here's an example to illustrate what I mean.

The wave theory of light is one of the greatest scientific advances. It hypothesized that light is a kind of wave motion in a medium (known as the ether), akin to water waves. The consequence can be deduced from this hypothesis that if an object is placed in the path of the light wave – such as a flat, round disk, for example – and we look closely at the shadow of the disk, we will see a white spot (this is formed by the light waves spilling around the edge of the disk and undergoing 'constructive interference' where the peaks of the wave reinforce each other giving a peak of intensity). When this white spot was observed, it was taken to be a significant confirmation of the hypothesis and the wave theory was regarded as justified.

The hypothetico-deductive account is a very well-known and much discussed view of how science works. It meshes with the Romantic view of discovery by insisting that science works by coming up with hypotheses in some creative way and then justifies these hypotheses by testing their experimental consequences. However, it has been subjected to the following criticisms:

1   There may be more to say about discovery than that it just
    involves a 'creative leap';

2   There may be more to experimental testing than just
    straightforward deduction.

We'll come back to the second issue in later chapters, but let's
consider the first in more detail.

Is creativity a myth?

The 'Romantic' view has been seen as pernicious and misleading.
Here's what Paul Feyerabend, a famous but quite radical philosopher
of science, said:

> The conceited view that some human beings, having the divine
> gift of creativity, can rebuild creation to fit their fantasies without
> consulting nature and without asking the rest of us, has not only
> led to tremendous social, ecological, and personal problems, it
> also has very doubtful credentials, scientifically speaking. We
> should re-examine it, making full use of the less belligerent forms
> of life it displaced.
>   (P. Feyerabend, 'Creativity – A Dangerous Myth', *Critical Inquiry*
> 13 (1987): 711)

Here, Feyerabend focuses on the social and broadly political consequences of this view of creativity and we might add to these the way
it encourages certain cranks to press their claims to have discovered
some new theory of quantum phenomena, or the true nature of
space and time, or just to have shown that Einstein was wrong (so
many of these people seem to have homed in on Einstein, whether
because of his stature as a physicist or because of some lingering
form of anti-Semitism). On a personal note, for a while it seemed to
me that just about every conference on philosophy of science or the
foundations of physics came with its own crank at the door – usually,
for some reason an engineer or computer scientist – eager to hand
out their poorly photocopied treatise. And quite often they would

preface their insistence that I read their work with the claim that 'it just came to me', 'I had this sudden flash of insight', 'it was my own private "Eureka" moment' and so on. These guys seemed to have found a way to sidestep or leap over entirely all the hard work that goes into a scientific discovery.

Now you might dismiss Feyerabend's comment as irrelevant: just because the social consequences of a particular view are unacceptable (to some) that doesn't mean that the view is false. We might just have to bite the bullet and accept these consequences. However, the last comment about the work involved is significant. We know that in many cases there is more to a discovery than just a sudden insight propelling you out of the bath.

Consider Archimedes, for example. He wasn't just some wild and crazy guy, sprinting through the streets with his new discovery (and little else). He was a brilliant mathematician and engineer who not only made a number of important 'theoretical' advances, but also designed and constructed war machines to defend Syracuse from the invading Roman forces, such as the 'Archimedes Claw', a massive grappling device designed to reach over the city walls and tip whole ships over, and parabolic mirrors which focused the sun's rays and set ships on fire (there are lots of websites covering Archimedes and his inventions; see, for example, https://en.wikipedia.org/wiki/Archimedes for a useful selection). Less militaristically, he also invented the 'Archimedes Screw', a device for raising water that is still used in some places today, and established a huge number of important mathematical results. Indeed, not only did Archimedes think his way through to these discoveries, he actually wrote a treatise called 'The Method' that set down how he arrived at certain results which prefigured the discovery of the calculus. The great tragedy, of course, is that while Archimedes and the Syracuse army were so concerned with defending their city from invasion by the sea, the Romans simply landed up the coast and attacked the 'back' way. Archimedes himself was found by a Roman soldier at the side of the road, busily sketching out some complicated geometrical result in the sand. When told to stop what he was doing and accompany the soldier, Archimedes replied, 'Don't disturb my circles!' and was promptly killed.

Likewise, although there is some evidence that Newton may well have been prompted into thinking about the earth possessing some attractive power by seeing an apple fall from a tree in his country garden (for a picture of the tree see: http://www.nationaltrust.org.uk/ woolsthorpe-manor/), he had already begun the development of the mathematics of the calculus, as well as performing important work in and lecturing on the theory of optics. In particular, his notebook from his early years of study at the University of Cambridge (before he had to leave to avoid the plague) reveals an active scientific and mathematical mind ranging over all kinds of problems and issues, including whether the 'rays' of gravity might be used to construct a perpetual motion machine (they can't, so don't get your hopes up!) and concluding that gravity must act on the interiors of bodies and not just their surfaces (see: http://en.wikipedia.org/ wiki/Quaestiones_quaedam_philosophicae#Gravity). In these early years Newton was clearly formulating not only the methodology and mathematics that he would use in his work but also getting a grip on the kinds of questions that needed to be answered.

And similarly, Mullis was no drug-loving Californian surfer dude, as he portrayed himself in his autobiography, or at least, he wasn't *just* a drug-loving surfer dude. He was also highly trained in biochemistry and had been working on specific problems in DNA replication for some time. Indeed, the basic principles behind PCR had been previously described in 1971 by Kjelle Kleppe, who earlier presented his work at a conference attended by (wait for it ...) one of Mullis's professors. Furthermore, his co-workers at Cetus disputed the claim that Mullis was solely responsible, arguing that it was much more of a team effort (something that Mullis himself fiercely rejected). What this illustrates is that the idea of the lone researcher boldly going where no scientist has gone before is perhaps a bit of a myth. Many commentators on science seem to perpetuate this myth but perhaps we may add to Feyerabend's negative consequences of the Romantic view, the distorted perception it fosters of the discovery process. The anthropologist Paul Rabinow examined Mullis's case and concluded that,

> Committees and science journalists like the idea of associating a unique idea with a unique person, the lone genius. PCR is, in fact, one of the classic examples of teamwork.
> (P. Rabinow, *Making PCR: A Story of Biotechnology*, Chicago: University of Chicago Press, 1996)

That there is more to discovery than simply a flash of insight is further illustrated by the case of Edward Jenner the discoverer of the smallpox vaccine. These days it may be hard to appreciate just how devastating smallpox was and how fearful people were of this disease that typically accounted for one in three of all kids' deaths and could wipe out 10 per cent of the population (20 per cent in cities where the infection could spread more easily). Following a ten-year campaign by the World Health Organization, the last recorded case of someone catching the disease by natural transmission occurred in 1977, and in 1980 the WHO declared 'Smallpox is dead!' (The last remaining samples of the smallpox virus are kept under massive security in two laboratories, one in the USA, the other in Russia. Some campaigners argue that these too should be destroyed, but scientists insist they should be retained for future study.) The basis of the eradication of this terrible disease – and also of numerous medical advances concerning other diseases – is the technique of vaccination, pioneered by Jenner.

There is a famous painting of Jenner, looking very much the country doctor, which you can see on-line at www.sc.edu/library/spcoll/nathist/jenner2.jpg. It shows him leaning nonchalantly against a tree, hat in hand, before an apparently innocuous and pleasantly bucolic scene, with countryside and cows and milkmaids in the background. But they're not in the painting because Jenner was standing outside in the farmyard when he was painted, or because the artist had a thing about milkmaids. They are there because the milkmaids are an integral part of the story of the discovery. Here's the story as it is often presented.

From the portrait, you might think that Jenner was merely a simple country doctor who had a particularly observant eye. And it is observation that is often put forward as playing a crucial role in this particular story. Jenner observed that milkmaids did not appear to catch smallpox as often as other people. Milkmaids and others who worked with cows often caught cowpox, a comparatively mild viral infection that could sometimes be caught by humans with only mild discomfort as a result. Jenner then made many observations of milkmaids over a period of almost four years and reached the hypothesis that inoculation with cowpox could protect against smallpox. One day during an outbreak of smallpox, he was consulted by Sarah Nelmes, a milkmaid who had a rash on her hand. Jenner assured her that this was a case of cow- rather than small-pox and, seizing the opportunity, took some of the pus from her sores and rubbed it into scratches on the arms of James Phipps, the eight-year-old son of his gardener. The boy felt some mild discomfort from the cowpox and was then injected with dried pus from a smallpox lesion: James survived and had effectively been vaccinated against smallpox (we might wonder what today's medical ethics committees would say about Jenner's approach!). Thus the technique of vaccination was discovered (from 'vacca', Latin for cow) and Jenner's hypothesis was confirmed. Not surprisingly, perhaps, there was some initial opposition, and there's a lovely cartoon from the period showing people in a doctor's surgery, having been vaccinated (you can see it at http://www.nlm.nih.gov/exhibition/smallpox/sp_vaccination.html), with cows' heads and hooves growing out of the arms and necks! But in 1853 vaccination was made compulsory by Act of Parliament and smallpox's days were numbered.

This suggests an alternative view of discovery, based on *observation*: we observe a number of relevant phenomena and, using that as a basis, come up with a hypothesis. The method by which we 'come up' with a hypothesis in this way is called *induction*.

The inductive account – a logic of discovery?

Let consider the simplest case, known as enumerative induction, which involves essentially observing increasing numbers of cases.

Imagine it's a bright sunny day and you're out walking in Roundhay Park, here in Leeds, and you notice a swan out on the lake and you observe that the swan is white. You look across the lake and you observe another swan and note that it is white too. Intrigued(!), you return the next day, when it's raining, and you make further observations of white swans and you then continue your observations on different days, under different weather conditions, from different vantage points around the lake. You then decide to extend your investigation and make similar observations further afield. So you

take the train to Alexandra Park in Manchester and there you observe
numerous other swans which all turn out to be white: and similarly in
St. James's Park, London, and, taking advantage of cheap rail travel
to Europe, the Luxembourg Gardens in Paris. Wherever you go, and
under all kinds of different conditions, the swans you have observed
are all white. Finally, given this observational basis you come to the
conclusion: all swans are white. We might put a fancy gloss on this
and state that you have induced the hypothesis, 'All swans are white.'
   We can represent this process a little more formally as follows:

Observations: Swan no. 1 is white (singular statement)
                    Swan no. 2 is white
                    . . .
                    Swan no. 666 is white
                    . . .
                    _____

Conclusion: All swans are white (universal statement)

There are a number of things to note about the above schema.
First of all, I have labelled the statements referring to the observations as 'singular' statements and the conclusion – the hypothesis
reached – as 'universal'. A singular statement is a statement about
something, an event say, which happens at a particular place, at a
particular time. A 'universal' statement, on the other hand, expresses
something that covers all places and all times. Scientific hypotheses,
and in particular those which express laws, are typically taken to be
universal in this sense; we shall return to this in a moment. So, the
general idea is that what induction does is take us from a number of
singular statements to a universal one. Compare this with *deduction*,
which forms the core of what we call logic. Here's an example of a
valid deductive argument:

Premise: All humans are mortal (universal)
Premise: Steven French is human (singular)

Conclusion: Steven French is mortal (singular)

Here we have gone from a universal statement (signified by the
prefix 'All') to a singular one (with the help of another singular

statement). With induction, it's the other way around – from the singular to the universal.

Which leads to our second point, which can be framed in the form of a question: How does induction work? You might think it is pretty obvious, particularly given the example of the white swans. After all, we've trudged round Europe observing hundreds, perhaps thousands, millions even, if we're feeling particularly obsessive, of swans and they've all been white, so what other conclusion can we draw? But there's still a bit of a leap involved: no matter how many swans we've observed, we haven't observed them all and it is to all swans that our hypothesis refers. Again, compare the case of deduction. In the case of the example above – which is simple, granted, but can be taken as representative – it is pretty clear how it 'works'. The conclusion is contained, in a sense, in the premises so that all we are doing is extracting it: if Steven French belongs to the class of humans (debatable perhaps but let's leave that for now) and if all humans are mortal, then Steven French has to be mortal. What you learn in your elementary logic classes are basically the rules and techniques for extracting the conclusions from various kinds of premises (there's more to deductive logic than that, but not much!). But in the case of the swans, the conclusion is most definitely *not* contained in the premises – it goes beyond them by referring to *all* swans. So, there's a bit of a mystery here.

And that it is a bit mysterious is further indicated by our third point: the conclusion of an inductive argument can be false. Consider: after trawling round Europe looking at swans, you decide to be a bit more adventurous and extend your net of observations even wider. So you take a plane and head off to Australia and there you observe the famed black swans of Queensland and your beautiful hypothesis comes crashing to the ground! (Actually, of course, you don't need to go all the way to Australia to observe black swans – there are even some in St. James Park!)

Bertrand Russell had another example which nicely makes the same point (if in a somewhat gruesome way, at least to us vegetarians): Out in the wilds of Norfolk there is a turkey farm and on that farm there is, contrary to all evidence from anyone who has ever had anything to do with domestic turkeys, a particularly bright young turkey, who observes that at 8 o'clock on a Monday morning he and his mates are all given a good feed. Not wishing to be hasty in drawing any conclusions, he keeps his counsel and continues to observe, noting that every day of the week he and his fellow turkeys are fed at 8 a.m. Again, not wishing to be precipitate in announcing his findings, he bides his time and continues to make his observations, noting that under different weather conditions, on warm days, cold days, rainy days and snowy days, he and the rest of the flock are fed at 8 a.m. And so, being a good inductivist, he assembles his set of singular statements and draws the conclusion that he and his fellow turkeys are always fed at 8 in the morning. Indeed, so confident is he of this hypothesis that when he announces it to his colleagues (who respond in typical turkey fashion, wandering around aimlessly and occasionally pecking at the ground and each other) he makes a prediction, that the next day they will all be fed at 8 o'clock. Unfortunately, however, the next day is Xmas (substitute Thanksgiving, if you're American)!

Let's again compare this situation with that of a deductive argument, using our previous example. In a valid deductive argument, if the premises are true then the conclusion must be true (that's what it means for the argument to be 'valid'). If you think of the conclusion as being extracted from the premises, you can see why this is so. But that is not the case for an inductive argument; no matter how good it is (and here we don't use the term 'valid'), no matter how many observations you have made, under however many different conditions, the truth of all the statements expressing these observations does not guarantee the truth of the conclusion. The observation statements, we recall, are all singular, whereas the conclusion is a universal one that goes beyond the set of singular statements, no matter how big that set is, so there will always remain the possibility that it will turn out to be false. Even if there were no black swans in Australia, there *could* be, somewhere else, perhaps, and you could never be certain your hypothesis was true, at least not in the way that you can be certain that the conclusion of a valid deductive argument is, if the premises are.

Now, one response might be to say that to be a swan is to be white, so that the conclusion has to be true. In other words, we might include 'having white feathers' in the definition of 'swan'. But then our 'conclusion' is uninteresting. Indeed, you might question in what sense it is a conclusion at all, since if a swan is simply defined to be a 'white bird of such and such a size, with such and such a shape of beak' and so on, then you really don't need to make any observations at all to 'discover' that all swans are white! The 'hypothesis' becomes true by definition, just like 'all bachelors are unmarried', to use a classic example from philosophy.

Alternatively, we might argue that although the conclusion of an inductive argument is not true, given enough observations, of sufficient variety, we can say that it is *probably* true. This sounds plausible: in the case of the white swans' hypothesis, the more swans you observe, under a greater variety of conditions, the more probable it is that the hypothesis is true. Indeed, this is how induction came to be treated by several of the great philosophers of science of the past hundred years or so. However, as plausible as it might seem initially, it doesn't take much thought to realize that it's actually quite a tricky project. How many observations do you have to make to raise the probability that the hypothesis is true by a certain amount? After all, there appear to be cases in which only one observation is needed: take the hypothesis 'fire burns' for example! And even after you've set the amount by which the probability is raised by each observation in cases such as that of the white swans, surely there comes a point of diminishing returns: after you've observed a million white swans, is the probability that 'All swans are white' is true going to be raised by the same amount if you observe another million swans? Surely not! And how do we factor in the different conditions? It's easy to see that the plausible idea soon becomes quite complicated.

But perhaps we should move on, since our discussion on how the probable truth of a hypothesis is supported and to what extent by our observations seems to have more to do with *justification* than discovery, and we'll come back to the former in later chapters.

Let's consider again the claim that discovery in science proceeds by making observations and using induction in some way. This sort of account certainly fits the view we noted in the introduction, that science is based on the 'facts', where 'based on' in this context is taken to mean 'discovered by means of' or something like that. (Of course, there are further questions we can ask, such as: What counts as a 'fact'? How secure are the 'facts'? We shall return to these in Chapters 5, 6 and 7.)

Now, how would we know that our claim is true? In other words, how do we know that scientific discoveries are made through observation plus some form of inductive argument? Well, you might say, it's obvious: we can look at the history of science and come up with case after case in which scientists arrived at their discoveries via observations. So the argument in support of our claim might go something like this:

Case 1:   Hypothesis 1 was discovered via observations
Case 2:   Hypothesis 2 was discovered via observations

...

Case 3, 478: Hypothesis 3,478 was discovered via
                 observations

Conclusion: All hypotheses are discovered via observations

Does this look familiar?! We appear to be supporting our claim that discovery proceeds by induction, using a form of inductive argument. Philosophers become rightly worried if a similar manoeuvre is used to justify induction as a good way to reason, since the manoeuvre only justifies the claim by using induction itself! And of course the conclusion of the above argument is no more guaranteed to be true than for any inductive argument. Indeed, it seems that we can easily come up with counter-examples to the claim that discovery is based on observations.

Let's return to the Jenner case. First of all, Jenner was not just a simple family doctor who made a series of observations, however astute. He was apprenticed to a surgeon at the age of thirteen and by twenty-one was the pupil of John Hunter in London, a renowned experimentalist and Fellow of the Royal Society (Jenner himself was eventually elected to the Society, not for his work against smallpox, but for his study of the life of the cuckoo! His claim that it is the cuckoo chick and not the adult bird that pushes the host bird's eggs out of the nest was only confirmed in the twentieth century with the advent of nature photography.) Furthermore, Jenner, like other doctors of the time, was completely familiar with the techniques of variolation and insufflation, whereby dried and ground-up pus from smallpox sores is injected into the skin or blown into the nose, respectively. These were techniques which came to Europe from China via Turkey and although they offered some protection, if virulent forms of the virus survived in the pus, the results could be fatal. Jenner himself was 'variolated' when he was a schoolboy and suffered so badly he never forgot the experience.

So, there was more going on in this case than a local GP making careful observations. In particular, there was a great deal of background knowledge about what might offer protection from smallpox and in particular about the effect of cowpox. Most importantly, perhaps, what we have here is not some mysterious inductive leap to a hypothesis after many observations but something more complex, in which observation certainly plays an important role, but other factors, such as background knowledge, also play a significant part. Indeed, there is a sense in which we can say that vaccination as a technique is a descendant of the cruder practice of variolation and Jenner's hypothesis builds on earlier – perhaps not clearly articulated – claims that inoculation with dried pus from smallpox sores protects against the disease. This idea, that discoveries in science are often not isolated events but can be situated in a given context and understood as following from previous work, is something we shall explore in the next chapter.

3

Heuristics

OK, let's review: we have put forward two views of discovery. One focuses on the so-called Eureka moment and both meshes nicely with the 'Romantic' view of creativity and supports what is known as the hypothetico-deductive view of science, according to which science works by coming up with hypotheses – How? We don't know and actually, as philosophers, we don't even care! – and then deducing experimental consequences from them which are then subject to experimental testing. And the other places the emphasis on observation and feeds into what we have called an 'inductive' account, whereby we amass a large number of observations collected under a variety of circumstances and somehow 'induce' a theory from that. Both views are inadequate.

So here is a third alternative, one which recognizes that scientific discovery is not just a matter of a creative leap, or some sort of mental lightbulb bursting into life, but nor is a matter of just slow and meticulous collection of observations. This is a view based on the idea that we can identify certain steps to discovery, certain moves that scientists make that are both rational and can play a role in explaining how discoveries are made and hence how science works; such moves come under the umbrella of what are known as 'heuristics'.

Heuristics: The steps to discovery

The word 'heuristics' is derived from the Greek word 'heurisko', meaning 'I find', and you probably won't be surprised to learn that it is related to the verb used by our friend Archimedes as he scrambled out of the bath. However, whereas 'eureka' has come to be associated with the flash of genius moment, heuristics is now understood to be the study of the methods and approaches that are used in discovery and problem solving. A heuristic lies somewhere between the stark formality of logic and the seemingly random and irrational flash of inspiration.

There has been a huge amount written on problem solving in the last fifty years or so. Much of it has drawn its inspiration from George Polya, who wrote a famous book entitled *How to Solve It* (Princeton: Princeton University Press, 1957). Polya was primarily concerned with solving mathematical problems and discovering proofs of mathematical theorems, and his general approach is perhaps not so earth shattering at first glance:

1 Understand the problem;

2 Make a plan;

3 Carry out the plan;

4 Review your work.

Understand the problem? Of course! Make a plan and carry it out? Who'd have thought it?! Brilliant! Well, perhaps I'm being a little facetious here. Polya did offer a set of heuristics which are more interesting, such as finding an analogy to the problem you're concerned with, and try to solve that, or, perhaps counter-intuitively, try to generalize your problem and solve the generalization, or, significantly as we shall see in a moment, try to find a problem that is related to yours which has already been solved. And, since Polya, the study of heuristics has expanded considerably. In psychology, for example, heuristic procedures are invoked to explain our everyday judgements that are typically made in the absence of complete information, or in complex situations and for which the usual rules of what is known as decision theory are inadequate. Certain of these rules embody the laws of probability theory and, by effectively flouting them, these 'everyday' heuristic procedures lead to what are known as 'cognitive biases'. So, it can be shown that by following these procedures:

1   People are often insensitive to the size of the sample they are considering and hence commit what is known as the 'base rate fallacy', for example;

2   They will violate certain of the laws of probability (such as that concerning the probability of two events occurring together) and, more generally;

3   People are prey to a range of biases associated with their personal familiarity with the event assessed.

As an example of 1), think about the following little conundrum:

There is a disease that one in a thousand people will get. Fortunately, there is a new test that has been developed for this disease *but* it is not completely accurate and in 5 out of every 100 cases it mistakenly indicates that the person has the disease when they actually don't; that is, the test has a false positive rate of 5 per cent. Now, you take the test and it comes up positive. Should you be worried? What is the chance that you actually have the disease?

If you think the chance is pretty high, perhaps as high as 95 per cent, then you're not alone. In one case study, involving sixty people given this problem, the answers ranged from 0.095 per cent to 99 per cent. Twenty-seven of the respondents gave the answer as 95 per cent; the average response was 56 per cent. Only eleven participants gave the correct answer, which can be calculated using probability theory, and which is 2 per cent. So this is not such a good test!

Of course, it does perform some diagnostic function, since the screening has increased the chance of picking out a diseased person by a factor of twenty. This is a big increase but obviously not as big as many people think. When considering whether a person who tested positive actually has the disease, people tend to ignore the information that only 1/1000 of the general population have it to begin with. This is known as the base rate and the fact that this base rate is small compared to the false positive rate leads to estimates that are way off the mark – dramatically so!

This is quite worrying and is even more so when you consider that the above question was not just asked of a group of 'lay' people but also of twenty fourth-year medical students, twenty residents and twenty attending physicians in hallway interviews at Harvard Medical School. Only four students, three residents and four attending physicians gave the correct answer. And there is evidence that because of this bias, medical staff have actually made incorrect diagnoses and recommended quite radical procedures in situations involving such tests, including, for example, early forms of mammograms.

Here is an example of 2). Consider the following scenario: Linda is thirty-one years old, single, outspoken and very bright. She majored in philosophy (of course!). As a student, she was deeply concerned with issues of discrimination and social justice and participated in anti-war demonstrations. Subjects were asked to rank the following statements by their probability, using 1 for the most probable and 8 for the least probable:

a) Linda is a teacher in elementary school.

b) Linda works in a bookstore and takes Yoga classes.

c) Linda is active in the feminist movement.

d) Linda is a psychiatric social worker.

e) Linda is a member of the League of Women Voters.

f) Linda is a bank clerk.

g) Linda is an insurance salesperson.

h) Linda is a bank clerk and is active in the feminist movement.

How would you rank statement h)? Would you rank it as more, or less, probable than statement f)?

It turns out that in the study, the average ranking of the joint claim 'Linda is a bank clerk and is active in the feminist movement' was higher than that of one of its conjuncts 'Linda is a bank clerk'. This violates what is known as the conjunction rule of probability theory, which states that the probability of two events happening together cannot be greater than that of either one by itself. And as in the 'base rate' problem, the Linda problem was given to three groups of subjects that differed in statistical sophistication: i) a statistically naive group of undergraduate students from the University of British

Columbia and Stanford University who had no background in probability or statistics; ii) a statistically aware group of graduate students in psychology, education and medicine who had taken several courses in statistics and were all familiar with the basic concepts of probability; iii) a statistically sophisticated group of graduate students in the decision science program at Stanford Business School who had all taken several advanced courses in probability and statistics. There were no statistically significant differences in responses among the three groups of subjects: that is, it did not appear to matter whether the subjects had taken advanced courses in probability or not.

What people appear to be using in these kinds of situations is an approach known as the so-called 'representativeness' heuristic, according to which the conclusions drawn are based on the expectation that a small sample will be highly representative of the parent population. This is the heuristic that, it has been suggested, lies behind the judgements reached in the Linda case, for example. So, it is claimed that,

> A person who follows this heuristic evaluates the probability of an uncertain event, or a sample, by the degree to which it is: i) similar in essential properties to its parent population; and ii) reflects the salient features of the process by which it is generated. Our thesis is that, in many situations an event A is judged more probable than an event B whenever A appears more representative than B. In other words, the ordering of events by their subjective probabilities coincides with their ordering by representativeness.
>
> (A. Tversky and D. Kahnemann, 'Judgments of and by Representativeness', in D. Kahnemann, P. Slovic and A. Tversky (eds), *Judgment Under Uncertainty: Heuristics and Biases*, Cambridge: Cambridge University Press, 1982, pp. 84–98)

So, rather than grinding through the theorem of probability theory, people considering statement h) above judge whether Linda being both a bank clerk and a feminist is more representative of someone with her background than her being a bank clerk alone. And hence, they evaluate the probabilities accordingly. Note that this 'representativeness' factor is established on the basis of some consideration of *similarity*; we shall return to this below.

Some believe these biases are ubiquitous and sound a note of warning:

> Since human judgment is indispensable for many problems of interest in our lives, the conflict between the intuitive concept of probability and the logical structure of this concept is troublesome. On the one hand, we cannot readily abandon the heuristics we use to assess uncertainty, because much of our world knowledge is tied to their operation. On the other hand, we cannot defy the laws of probability, because they capture important truths about the world.... Our problem is to retain what is useful and valid in intuitive judgment while correcting the errors and biases to which it is prone.
>
> (Tversky and Kahnemann, ibid., p. 98)

There has been a great deal of commentary on these conclusions, on whether the biases are as widespread as the above commentators claim and whether they extend to scientific reasoning, for example. One response has been to suggest that what is going on in these studies is that the subjects, and by implication most of us when it comes to 'everyday' reasoning, are using certain non-standard models or representations (non-standard in the sense that they do not conform to the laws of probability theory, say). It has been argued that, for example, in those 'natural' judgements for which subjects fail to take into account the relation between sample size and sampling error, they are simply not incorporating this relation into the intuitive model they set up to handle the problem. Instead, they employ the 'representativeness' heuristic, based on the expectation that a small sample will be highly representative of the parent population. It has been suggested that this particular heuristic procedure can actually be split into two: a model construction heuristic, according to which there should be a close match between the underlying model and the structural features of the data, and a separate heuristic for judging likelihood of outcomes, which suggests that an outcome is more likely if its structure is more *similar* to that of the assumed underlying model. And, again, subjects ignore the base rate, for example, because they simply do not have to hand the appropriate models which would allow them to process base-rate information (see R. N. Giere, *Explaining Science*, Chicago: University of Chicago Press, 1988, p. 173).

Hence the difference between a lay-person using these heuristics and someone who knows and applies decision theory or the laws of probability is simply that,

> The latter have at their disposal a whole family of probabilistic models they can deploy in handling the sorts of problems presented. And, as a matter of fact, rather than of logic, these models are better fitted to the kinds of cases at issue.
>
> (Giere, ibid.)

Now, we are not suggesting that scientists, in trying to come up with new hypotheses or theories, use these particular kinds of heuristics. Nevertheless, they do use something similar, which has the same kind of character. In particular, the domain of scientific discovery is structured in a certain way, in the sense that it is not simply marked out by the Eureka moment or flash of genius, but incorporates certain moves and approaches that take scientists to where they need to go.

Of course, to say that discovery is 'structured' is not to say that it follows an algorithmic procedure, in the sense that there is a set of rules and all that needs to be done is to apply them in order to obtain a new theory. But even if we can't have a 'logic' of discovery, we can still talk of a rationale which can be identified and described by examining appropriate case studies. If that is so, you might ask, what about creativity? Perhaps the line between creativity and heuristics can be drawn through psychology. On one side, we have the private – and perhaps subconscious – circumstances, which lead a scientist to an idea, whereas on the other we have the connections between that idea and the relevant context, which are covered by the heuristic procedures I will sketch out below. The field of creativity – how it is that scientists actually come up with the ideas they do – is concerned, at least in part, with the former circumstances. Nevertheless, a great deal of what is put down to 'genius' and 'creativity' can be understood as the judicious perception and exploitation of a particular heuristic situation.

The overall idea here, then, is that discovery is more complex than is suggested by either of the two views discussed in the previous chapter, but nevertheless, we can identify certain moves at both the 'experimental' and 'theoretical' levels. Let's begin with a very simple example at the experimental level.

Experimental: Observing similarities between phenomena

Let's consider, as an example, the explanation of lightning as an electrical discharge, an explanation proposed in the eighteenth century. Cutting a long and interesting history short, the invention of electrical machines – such as friction machines, in which a glass sphere or cylinder was rubbed by a rotating pad, leading to a build-up of static electricity (a useful website can be found at http://www.sparkmuseum.com/FRICTION_HIST.HTM) – led to the observation that there was a similarity between the sparks generated by such machines and lightning. In 1749, the famous scientist, politician and polemicist, Benjamin Franklin, noted the relevant points of similarity:

1 Giving light
2 Colour of light
3 Crooked direction
4 Swift motion
5 Conducted by metal
6 Crack or noise in exploding
7 Subsisting in water or ice
8 Rending bodies it passes through
9 Destroying animals
10 Melting metals
11 Firing inflammable substances
12 Sulphurous smell

These observed similarities led Franklin to put forward the hypothesis that lightning was nothing more than a form of electrical discharge. Here we see an apparently simple form of discovery, based on the heuristic move of noting certain similarities between the relevant phenomena. (Of course, this cannot be the whole story, as it raises the further question of how we decide which phenomena are relevant!)

Jumping ahead to the subject of subsequent chapters, Franklin's hypothesis was then tested.

Test 1: Dalibard and his 40-foot rod

Across the Atlantic, in Paris, Thomas-François Dalibard (a French scientist and friend of Franklin) constructed a 40-foot metal rod designed to 'draw lightning down' (it was apparently grounded using, appropriately enough, wine bottles). An 'old dragoon' was then instructed to approach with an insulated brass rod (it's not clear why Dalibard himself didn't perform this part of the test; perhaps he wanted to keep his distance in order to make the necessary observations, or perhaps he just wanted to keep his distance!). Following a lightning strike on the rod, there was an 'infernal' flame and odour, causing the dragoon to run away terrified and call for the local priest, who drew sparks from the rod. Following this striking demonstration, Dalibard announced that 'Franklin's idea ceases to be a conjecture. Here it has become a reality.' The next test is better known.

Test 2: Franklin flies a kite

Many of us are familiar with this story, or may even have seen drawings or paintings of Franklin flying a kite with a metal key attached, in order to draw the lightning down and observe the sparks produced. Let's consider this episode in a little more detail.

Having published his hypothesis that lightning was simply a form of electricity, Franklin also described how it might be tested by observing the discharge produced when lightning struck an elevated metal object, such as a pole. His first thought was that a church spire might do the trick, but while waiting for the erection of the spire on Christ Church in Philadelphia, it occurred to him that he could attain what he thought would be the necessary height by using a common kite. In the absence of kite shops in eighteenth-century Philadelphia, Franklin set about (in Blue Peter mode) making one from a large silk handkerchief and two cross-sticks of appropriate length. He then had to wait for the next thunderstorm, but as soon as one was observed approaching he set off across the fields, where there was a shed he could use to store his equipment etc. However, fearing ridicule if his test should fail, Franklin told only his son, often portrayed in pictures of the scene as a child, but actually a twenty-one-year-old, who helped him get the kite up.

Here's an account of what happened (kids, don't try this at home):

> The kite being raised, a considerable time elapsed before there was any appearance of its being electrified. One very promising cloud had passed over it without any effect; when, at length, just as he was beginning to despair of his contrivance, he observed some loose threads of the hempen string to stand erect, and to avoid one another, just as if they had been suspended on a common conductor. Struck with this promising appearance, he immediately presented his knuckle to the key, and (let the reader judge of the exquisite pleasure he must have felt at that moment) the discovery was complete. He perceived a very evident electric spark. Others succeeded, even before the string was wet, so as to put the matter past all dispute, and when the rain had wet the string he collected electric fire very copiously. This happened in June 1752, a month after the electricians in France had verified the same theory, but before he heard of anything they had done.
>
> (J. Priestley, *The History and Present State of Electricity*, with original experiments, 1775, Vol. I, pp. 216–17)

'Exquisite pleasure' indeed! There are a couple of things worth noting about this episode and we shall come back to them in later chapters. First of all, note that the experiment initially seemed to have been a failure. According to certain views that we shall examine in some detail in the next chapter, Franklin should perhaps have concluded that his hypothesis was false. However, he didn't, but observed more closely and in particular noted that when the string was wet, he got a very noticeable result (water being a good electrical conductor). Sometimes the conditions have to be right to get the best result, or any result at all, which means if we don't observe what we expect, it may be that the conditions aren't right, rather than the hypothesis is at fault.

Moving up the structure as it were, the observation of similarities at both experimental and theoretical levels has also acted as a powerful heuristic move in the discovery of new theories.

Experimental/theoretical: Similarities and unification

Here we see how similarities can be tracked all the way up from the level of phenomena to the highest theoretical level. In 1819, the Danish scientist Hans Christian Øersted discovered that when a magnetic needle was brought close to a wire carrying an electric current it was deflected. The French physicist André-Marie Ampère (who gave his name to the unit of current) showed that current carrying wires could act as magnets and his compatriot François Arago used such a wire to magnetize a chunk of iron. These observations all suggested a close association between electricity and magnetism and in 1831, Michael Faraday in the UK and Joseph Henry in the USA, independently discovered that moving a magnet close to a wire could induce an electric current in it. Faraday also introduced the idea that electricity and magnetism produced their effects via lines of force propagating across space and this led to the idea of electric and magnetic fields (I'm compressing a huge amount of history here!).

James Clerk Maxwell, one of the scientific giants of the nineteenth century, then unified these two fields of study by developing a new set of laws of electromagnetism, embodied in his famous set of equations. According to these laws, just as changing magnetic fields produce electric fields, so changing electric fields create magnetic ones. Maxwell then surmised that electric and magnetic fields oscillating at right angles to each other would persist across space. When he calculated the speed at which this electromagnetic field would travel, Maxwell found that it was equal to the speed of light, leading to the suggestion that light was an electromagnetic wave. In 1884, Heinrich Hertz (another great scientist who gave his name to one of the essential units of modern life!) reformulated Maxwell's equations, fully revealing the fundamental symmetry between electricity and magnetism. Four years later, he constructed experiments to confirm one of the predictions of Maxwell's theory, namely that other electromagnetic waves such as radio waves should also travel at the speed of light. We shall return to consider Hertz's observations in a later chapter.

What we have in this little encapsulated history is a series of similarities noted at both observational levels and theoretical levels that drove, first, the unification of electricity and magnetism and subsequently the identification of light with an electromagnetic field. Noting such higher-level theoretical similarities has also become tremendously important for twentieth-century science. Of course, such developments crucially depend on what is being regarded as similar with what – it has famously been said that anything can be made to seem similar to anything else! By formulating theories in certain ways, so that certain mathematical features of the equations become apparent, ostensibly different theories can be seen to be similar in terms of their symmetry properties (if an object looks exactly the same when reflected in a mirror, it is said to be symmetric under reflections; the kinds of properties referred to here are like this but expressed in terms of higher-level mathematics). It was on this basis that Weinberg (he of the red Camaro), Abdus Salam and Sheldon Lee Glashow noted that certain similarities held between electromagnetism (in its modern, quantum form, known as quantum electrodynamics) and the weak nuclear force responsible for radioactive decay. This new unified theory, known snappily enough as the electroweak theory, predicted the existence of three new particles in nature and their subsequent discovery in 1983 was hailed as a significant confirmation of the theory (with Nobel Prizes all round for Glashow, Salam and Weinberg). And so it goes. The next step was to achieve a similar unification with the strong nuclear force responsible for holding the nucleus together, and that just leaves gravitation, but bringing that force into the synthesis turns out to be a different proposition altogether.

Theoretical: Correspondence

In general, theories do not just pop into the head of a scientist, as the Romantic view would have us believe; nor do they typically emerge, inductively or otherwise, from observations, no matter how many we make and under whatever diverse conditions. As I indicated in the cases of Archimedes and Mullis, the ground is usually well prepared and the scientist calls upon a range of background knowledge as the relevant context in which to formulate the new hypothesis. But it can be argued that we can make an even stronger claim than that and insist that in many cases new theories are built upon the backs of the old. The idea is that scientific progress is an essentially cumulative process and new theories are constructed from those that have gone before. Now this is a contentious thesis about scientific progress since it appears to conflict with what we know from the history of science, namely that science sometimes undergoes quite radical change. These changes are often called 'revolutions' and hence we have the 'Quantum Revolution' of the early twentieth century, the 'Einsteinian Revolution' in which relativity theory was introduced, the grand-daddy of them all, the 'Scientific Revolution' of the seventeenth century, associated with the likes of Newton, and even the 'DNA Revolution' in biology. How can scientific progress be essentially a cumulative affair, with new theories somehow built up from their predecessors, in the face of these revolutions?

Some have argued that it cannot. Thomas S. Kuhn, for example, in a work that has had a huge influence, going beyond the philosophy of science, insisted that scientific revolutions are marked by sharp, dramatic breaks in which not only do theories change, but also what counts as a 'fact' and even scientific methodology. The title of his most famous book, *The Structure of Scientific Revolutions*, indicates the central focus: once a particular scientific field – physics, psychology, microbiology, whatever – becomes sufficiently organized so that there is widespread agreement on what the central problems are, how they should be tackled, what would count as a solution to them and so on, it can be described as adhering to a certain 'paradigm' or 'disciplinary matrix'. This sets down the rules of the game as it were, in the above terms of determining the central problems of the field, the methodology to be used when addressing them, the criteria for determining when they have been solved and so on. New workers in the field are induced into the paradigm through their education and training and, within the paradigm/disciplinary matrix, what Kuhn called 'normal science' is conducted, with the central focus on problem solving.

Those problems that cannot be solved within the framework of the paradigm are set aside as anomalies. Gradually these anomalies accumulate until a point is reached when someone – typically a younger scientist with little to lose! – declares that the old paradigm is bankrupt and sets about constructing a new one. Typically, insisted Kuhn, in order to solve the anomalies, we need not only a new theory but a new way of doing things, so that what appeared to be facts in need of explanation according to the old paradigm are dismissed according to the new one. So, consider the transition from Aristotelian physics to Newtonian: according to the former, when something is given a push, we need to explain why it continues to move (ignoring friction or air resistance). That generated theories in which moving objects are continually pushed from behind as it were. According to Newton, however, that an object set in motion continues to move needs no explanation at all; what has to be explained are *changes* to that motion, through the effects of forces. Furthermore, because the new framework is so different, because what counts as a fact, what needs to be explained, what counts as an explanation, how theories are justified etc. are not the same in the new paradigm as they were in the old, we cannot compare the new and old paradigms at all – they are, according to Kuhn, 'incommensurable', because there is no common basis for comparison.

Now that is quite a radical claim and one that appears to undermine the very basis of scientific progress in any meaningful sense; after all, how can you say that there has been progress from one theory to another across a revolutionary divide if there is no common framework in terms of which the two can be compared. However, Kuhn himself backed off from this claim in the later edition of his book, and suggested that the standard measures of comparison, such as simplicity, empirical support etc., could still be applied, although he retained doubts about a cumulative sense of progress.

And others have argued that if we look closely at so-called scientific revolutions we can in fact discover sufficient common-alities across the divide to be able to say that not only has there been progress, but we can discern how subsequent theories were constructed on the basis of their predecessors. This idea is enshrined in something called, rather grandly, 'The General Correspondence Principle'. Roughly speaking this states that any acceptable new theory should account for its predecessor by 'degenerating' into that theory under those conditions under which the predecessor has been well confirmed by experimental tests (a good account of this notion can be found in H. R. Post (1971 [1993], 'Correspondence, Invariance and Heuristics', *Studies in History and Philosophy of Science* 2: 213–55; reprinted in S. French and H. Kamminga (eds), *Correspondence, Invariance and Heuristics: Essays in Honour Of Heinz Post*, Boston Studies in the Philosophy of Science 148, Dordrecht: Kluwer Academic Press, 1993, pp. 1–44). A pithy way of putting this is to say that we always keep the best of what we have. And a nice example would be the periodic system of the elements, which survived the Quantum Revolution.

Now we'll come back to this idea when we consider the issue of adopting a realist stance towards our theories in Chapter 8, but what has it got to do with discovery? Well, the idea is as follows: if it is the case that although some things change, at a high level, say, but a lot remains the same at the lower levels, and new theories are built on and retain these same elements, then another heuristic move would be to focus on these elements and construct your new theory on that basis. The trick, of course, is to identify which parts of the old theory should be kept when you construct the new one – get that right and you can help yourself to a Nobel Prize!

Theoretical: Flaws and footprints

Another fairly obvious heuristic manoeuvre is to head in the opposite direction, as it were, and look for flaws in a theory and see if rectifying them can lead to a new, better theory. In this way, such flaws can be seen as 'footprints' of the new theory.

A pretty major flaw is if a theory is internally inconsistent and typically this would be enough to rule any such theory out of contention right away; indeed, inconsistent theories generally don't make it much beyond the initial thoughts of their discoverers and certainly not into the scientific journals. However, a famous example is that of Niels Bohr's theory of the atom, which hypothesized that electrons of the atom orbit the nucleus and can only hop up from a lower orbit to a higher one if energy is taken in, or drop down from a higher orbit to a lower one if energy is given out. The energy taken in or given out, respectively, has to be equal to the difference in the energies of the orbits and Bohr applied the new quantum theory of Max Planck to show how those energies were equivalent to certain quanta of radiation. This allowed him to explain the spectra of radiation given off when different elements are heated and, more precisely, to explain why those spectra contained discrete lines (such lines corresponding to electrons hopping between different orbits). However, although this was the first quantum theory of the atom, it incorporated the fundamental principles of pre-quantum, classical physics, one of which states that anything moving in a circle, such as the moon orbiting the earth, or electrons orbiting a nucleus, is undergoing acceleration and charged particles that accelerate give off energy (this is basically how TV and radios work – signals are generated by causing electrons to accelerate in certain ways). So, according to this theory, electrons orbiting an atomic nucleus should be radiating and giving off energy continuously: indeed, they would very quickly lose energy and spiral into the nucleus, resulting in no atoms, and no Bohr to come up with the theory! Bohr simply insisted that in his theory electrons in orbits did not radiate: they did so only when they changed orbits. But he didn't explain how this could be reconciled with the classical principles he invoked and hence there appeared to be an inconsistency in his theory: on the one hand, he wanted to use certain principles, and on the other hand, he wanted to deny them, or certain aspects of them.

Now explicating how Bohr was able to do this and still come up with a meaningful explanation of the spectra given off by different elements is a complex story but this major flaw was one of the factors that drove scientists to come up with a better, more complete quantum theory of the atom, now known as quantum mechanics.

Theoretical: Taking models seriously

Another heuristic move is to construct a model of the system or process of interest, rather than a complete theory, and then take the model seriously as representing at least some aspects of the system or process accurately. We'll be discussing models again in Chapter 7 but often models are built because the full theory would be too complex to work with, so idealizations are introduced which allow scientists to produce meaningful results with limited resources. Consider that staple of introductory physics courses, the simple pendulum. In practice you would construct one of these in the lab by tying a lead weight to a piece of string and then tying the string to a clamp stand or something similar, before setting the weight swinging and measuring the way the period of the swing changes with the length of the string, for example. Now, when you come to represent this situation in order to write down the relevant equations, you generally don't take into account the effects of the friction between the string and the clamp stand, or the effects of air resistance; furthermore, if you're considering the standard formula which expresses the relationship between the period and length of string, you only consider setting the weight swinging at small angles from the vertical, because at larger angles the formula breaks down. What you are doing here is constructing a simplified model of the situation that allows you to obtain reasonably accurate results, reasonably straightforwardly. Of course, the trick to model building is not to idealize too much, else the model won't represent the situation at all!

Here's another 'classic' example: the billiard ball model of a gas. Constructing a decent theory of how gases behave is an incredibly difficult business because not only do you have millions upon millions of atoms, but they're all moving in different directions, colliding with each other and the walls of the container and exerting forces on each other and the walls. One way of beginning to get a handle on representing this situation is to assume the atoms are incredibly hard, like billiard balls, or pool balls, so when they collide they bounce right off each other (the collisions are then described as 'inelastic') and there are no long-range forces affecting their motion. This turned out to be such a productive way of conceiving of a gas that it's become a standard example of a scientific model.

Here a physical system – the gas atoms – is represented in the model by something else – the billiard balls. Another example is the so-called 'liquid drop' model of the atomic nucleus. In this case the nucleus is represented as a drop of liquid and just as such a drop vibrates and shakes and splits apart when energy is pumped into it, so the atomic nucleus undergoes fission when energy – in the form of sub-atomic particles for example – is similarly pumped into it. These kinds of models in which one thing – the physical system of interest – is represented in terms of another are called 'analogous models', because the billiard balls and liquid drop are said to be analogies of the gas atoms and nucleus, respectively. Crucially, in these cases the analogy is with something familiar to us (or to those of us who misspent our youths) such as billiard balls or drops of liquid.

But how can such models be used as heuristic devices in discovery? Consider this passage from a textbook on nuclear physics, which describes the 'method of nuclear models':

> This method consists of looking around for a physical system, the 'model', with which we are familiar and which in some of its properties resembles a nucleus. The physics of the model are then investigated and it is hoped that any properties thus discovered will also be properties of the nucleus. ... In this way the nucleus has been treated 'as if' it were a gas, a liquid drop, an atom and several other things.
>
> (L. R. B. Elton, *Nuclear Sizes*, Oxford: Oxford University Press, 1961, p. 104)

So the idea is that we establish the model on the basis of some form of correspondence between some of the properties of the elements of the system and some of the properties of the object or set of objects in terms of which we're modelling the system. In a famous study of models and analogies, Mary Hesse called this the 'positive analogy'. Now of course there are some properties which feature in the model but do not represent properties of the system we're modelling: billiard balls may be coloured, for example, and pool balls have numbers but gas atoms have neither. Such properties constitute what Hesse called the 'negative analogy'. But then crucially, there are properties which feature in our model but which we're not sure whether or not they're possessed by the system we're studying; Hesse took such properties to form the 'neutral analogy' and it is here where all the action is as far as discovery is concerned, because it is by exploring the neutral analogy and determining whether properties in the model hold in the system that we discover new features of the system.

Consider a drop of liquid again. When a liquid changes phase and becomes a vapour, heat is taken in, known as latent heat. This latent heat is a classic indicator of a phase change and is independent of the size of the liquid drops. This arises because of the short-range nature of the inter-molecular forces, which 'saturate', in the sense that once enough close neighbours have been bound the presence or absence of more distant molecules does not alter the binding of a given one. This implies that the total energy is proportional to the total number of particles in the system, since each particle makes a fixed contribution to this energy. Pursuing the analogy between a liquid drop and the nucleus then leads to the suggestion that the nuclear forces also saturate in this way, which accords with observed experimental results: the binding energy per nucleon is approximately constant over a wide range of nuclei. So here we see how exploring the neutral analogy can help us understand the properties of the system being modelled.

As I said, a model can be physical or conceptual. A famous example of the former is Crick and Watson's model of DNA, revealing its double helix structure by means of a model built out of wire and steel plates (a photo of the pair proudly standing beside their model can be found on the website of the Science Photo Library, at http://www.sciencephoto.com/; and a close-up of the actual model itself can be seen at http://www.sciencemuseum.org.uk/on-line/treasure/objects/1977–310.asp). This is one of the major discoveries of the twentieth century, of course, and it is nicely discussed in some detail in Robert Olby's book, *The Path to the Double Helix* (London: Macmillan, 1974, 1994). The history (which I can only crudely outline here) nicely illustrates the role and importance of background knowledge, on the basis of which we can begin to discern the correspondences between the 'old' and 'new' theories. So, the initial problem was to account for the transmission of genetic information, and through a combination of experimental and theoretical work this was identified with deoxyribonucleic acid, and not a form of protein as originally thought. The next problem was to work out the structure of DNA. Now one of my University's claims to fame is that it was here at Leeds that William Astbury developed the techniques for producing x-ray diffraction patterns of fibres, proteins and, finally, DNA itself (for details see: http://www.leeds.ac.uk/heritage/Astbury/). This built on the work of the great physicist Bragg, who was also at Leeds for a time, and who showed that by studying the diffraction patterns produced by the scattering of x-rays off crystals, one could determine the structure of these crystals. Astbury built on this work to show that DNA had a regular structure, although his data was too rudimentary to determine what it was.

This data was improved upon by Maurice Wilkins and Rosalind Franklin, and the latter in particular applied her knowledge of x-ray diffraction techniques to produce a series of detailed diffraction pictures of DNA. Linus Pauling also combined x-ray diffraction patterns together with attempts to model the relevant structures and discovered that many proteins included helical shapes. His attempt to similarly model the structure of DNA was unsuccessful, and Franklin rejected modelling as a mode of discovery completely, stating that one should only construct a model after the structure was known (from the ground up as it were, using the x-ray studies).

Francis Crick and James Watson had no such reservations and using Franklin's data (controversially, since the crucial x-ray pattern was shown to them without her permission), together with their understanding of the biological and physical constraints, produced a physical model, built out of wire and tin, which explained the data presented in Franklin's diffraction patterns. By positing a double-helix structure, it also explained how genetic information could be transmitted, through the separation of the two intertwined strands and the replication of DNA by creating the complement of each strand (their famous paper can be found at www.nature.com/nature/dna50/archive.html). This proposed replication mechanism was subsequently confirmed experimentally and Crick, Watson and Wilkins received the Nobel Prize in 1962 (unfortunately, Franklin did not, as she had died from cancer and Nobel Prizes are never awarded posthumously).

This, of course, lies behind Mullis's subsequent discovery of the polymerase chain reaction and his Nobel Prize, but, as in his case, the discovery did not take place in a 'flash of genius' or Eureka moment. Using just this potted history, we can pull out three intertwined(!) strands in the discovery:

Background knowledge Any proposed structure for DNA had to account for the transmission of genetic information. There was already quite a considerable body of experimental and theoretical work, on the basis of which a proposal for this structure could be built. Indeed, if we probe deeply enough, we can discern interesting commonalities and correspondences between these earlier attempts and Crick and Watson's discovery.

Experiment Wilkins and Franklin used x-ray crystallography to establish that DNA had a regular, crystalline structure. Again, this built on earlier work and involved an interesting mix of experimental skill and theoretical knowledge.

Theoretical/model building Pauling discovered the basic helical structure of the protein molecule by building models to fit the experimental facts; Crick and Watson then applied the same model building techniques to discover the structure of DNA.

Combining these three strands, Crick and Watson built *models*, constrained by *experimental results* and the requirement that the structure be such as to allow *transmission of genetic information*. And what is important to note from the perspective of this chapter is that, first of all, this discovery was not wholly irrational, or mere guesswork, as the hypothetico-deductive account might claim, but second, neither was it made via induction from observations: rather it involved a complex combination of different heuristic moves.

That's it for discovery. In the next chapter, we'll consider the role these hypotheses, models and theories might play in scientific explanations and then we'll see what happens when we throw them to the wolves of experience.

Exercise 1

Chapters 2 and 3

Here are some questions to encourage you to think further about scientific discovery:

Q1 Do you think Kant's view of creativity is correct? Can you think of any counter-examples (from art, science, wherever …)?

Q2 What is the 'Eureka view' of discovery? What are its pros and cons?

Q3 What is the hypothetico-deductive account of science? How does discovery fit into it? What are its pros and cons?

Q4 Explain the distinction between the contexts of discovery and justification. Do you agree that philosophy should only be concerned with justification and not discovery?

Now watch the video (it's only about seven minutes long) available at: http://www.ted.com/talks/how_simple_ideas_lead_to_scientific_discoveries.html

Q5 What's the point about Feynman, the ball and the wagon?

Q6 Do either of the discoveries discussed in the video (of Eratosthenes and Fizeau) fit any of the accounts of discovery covered here? If not, can you come up with an alternative account that they do fit?

Here are some more advanced questions that might be interesting to think about:

A Give an example of a 'Eureka moment', other than those from the readings or the slides. Argue that your example is a genuine 'Eureka moment'.

B   Take one of the examples discussed in this book. Argue against the idea that it constitutes a genuine 'Eureka moment'.

C   What should we make of the problem of induction? Is it *particularly* a problem regarding scientific knowledge claims?

D   Give an example where induction has really worked in science.

E   Can induction itself be enough to yield a novel scientific discovery? Come up with an example in which the hypothetico-deductive model fails and state the problem(s) it faces.

F   Illustrate the context of discovery and the context of justification in relation to a specific scientific theory.

4

Explanation

Introduction

So, you've made your discovery, you think you're on to something interesting but now people are starting to ask, 'Where's the evidence?' That question takes us into the next phase of how science works, which has to do with how theories, hypotheses and models relate to the phenomena and how the latter, understood as evidence, support – or not – these theories, hypotheses and models (we'll look at the notion of a 'phenomenon' in a little more detail in Chapter 7). This is what philosophers of science call 'justification' (of course, whether the discovery and justification phases can be cleanly separated is itself a philosophical question).

One way in which theories etc. relate to phenomena is by *explaining* them and, indeed, much of the power of science would seem to lie in its ability to explain things. A quick scan of the BBC's science webpages (http://www.bbc.co.uk/science/0/) reveals a wide range of explanations including why honey bee populations have declined so dramatically in recent years (pesticides, a virus and changing land use), why there are hot geysers on one of Saturn's frozen moons (the gravitational pull of Saturn itself), why certain male finches are 'losers' in courtship (failure to socialize with females during adolescence: happens to us all ...), why Usain Bolt (for those of you reading this in the distant future – he was an Olympic sprinter) can run so fast (extraordinarily large stride length) and so on. These explanations were prompted by very different kinds of projects, some practical, some theoretical, some urgent and some undertaken just because the phenomena involved seemed interesting.

The job of the philosopher of science is to ask a more general set of questions: What is the nature of scientific explanation? What is its form? Does one form fit all, or do different sciences use different kinds of explanation? And does explanation always lead to under-standing? In this chapter we'll take a closer look at these questions and outline some answers.

So, consider rainbows (he suggests, looking out on a grey and rainy day): the great mathematician, scientist and philosopher René Descartes is usually credited with giving the first scientific expla-nation of this phenomenon, drawing on an emerging understanding of the theory of light. Without going into all the details, the expla-nation runs as follows: given the laws of refraction and reflection of light rays, if you stand facing the rain (or mist or even a waterfall) but with the sun behind you (or even the moon, if it's bright enough) and usually at a low angle (no higher than 42 degrees, unless the observer is on a plane or on top of a mountain) you will see a rainbow (or mistbow, or waterfallbow ...).

Now although this explained why we see rainbows where they are in the sky, Descartes couldn't explain why we see the colours we do. It took that other great mathematician and scientist, Isaac Newton, to explain that feature of the phenomenon, using Descartes' theory as his basis. But even then, there are other features that Newton could not explain (like the violet and green arcs that you can see – if you're sharp-eyed enough – underneath the main bands of colours, as well as other neat phenomena, such as double rainbows, 'stacker' rainbows, supernumerary rainbows and so on) and in fact the expla-nation of all the various features of rainbows turns out to be quite a complicated affair, involving a lot more maths and physics than I could possibly go into here (for some indication of what's involved as well as some pretty pictures, see: http://en.wikipedia.org/wiki/Rainbow). Indeed, and perhaps surprisingly for a phenomenon so common, research on rainbows is on-going, using quite advanced mathematics and computer simulations. That's true of many explanations in science but we can use Descartes' account above as representative of a general philosophical framework for understanding explanations in science, known as the 'deductive-nomological' view (or D-N for short).

The deductive-nomological view of explanation

The name sounds more complicated than it really is! Recall in Chapter 2 how we discussed the 'hypothetico-deductive' view of how science works: scientists begin with hypotheses and then deduce observational consequences from them. Well, the D-N view of explanation is similar: reading the name from right to left, the word 'nomological' means relating to laws (it's from the Greek 'nomos' for law), so the D-N view tells us that when we explain something, we begin with the relevant laws that are expressed in our theory or hypothesis and then we *deduce* from those laws a statement describing the relevant feature of the phenomenon that we want to explain (in this case why rainbows form where they do).

That sounds straightforward but if you've been paying attention you'll have noticed that Descartes' explanation involved more than just the laws of optics – it included some quite specific conditions that have to be met in order to observe a rainbow: namely that the sun has to be behind you, the drops of rain/mist/waterfall have to be in front and you have to look at an angle of 42 degrees. These kinds of conditions are crucial for the explanation to work (some of them are called 'initial conditions' in scientific contexts, while others are 'boundary conditions') as they effectively set the context in which the relevant laws can be applied. So, the general form of the explanation would be: from the laws of optics and with the sun behind you, the rain ahead and looking at an angle of 42 degrees, you can deduce that a rainbow will be observed. Even more generally, we can represent this kind of view of explanation like this:

Laws + Context (initial conditions etc.)
↓
Phenomena

Prediction and explanation

Notice that I said 'a rainbow *will* be observed' in the above description. Another significant feature of science is that it *predicts* things and one of the virtues of the D-N view is that it does double duty as an account of both explanation and prediction.

So, consider the following (here given only in broad and sketchy terms): On the basis of Newton's laws and observed irregularities in the orbit of Uranus (no sniggering now), astronomers were able to predict the existence of a new planet, Neptune (that in fact may have been observed but not recognized as a new planet earlier); using the theory of evolution, together with already known fossils, biologists were able to predict that a 'transitional' fossil between fishes and the first four-limbed vertebrates would be found in rocks of a certain age and associated with a certain habitat (and indeed, when the rocks of that kind were examined, the fossils were found); and more recently, using the laws of elementary particle physics (embodied in the so-called Standard Model), together with certain data, physicists were able to predict the existence and, crucially, properties of a particle called the 'Higgs boson', aka the 'God particle' because it basically gives mass to those particles that have it (and again, when they looked for the Higgs, they found it where it should be – Nobel Prizes all round once more!).

Let me give one more example because it played an important role, not only in the history of science, but in the history of the philosophy of science, because I'll come back to it again in later chapters and because it's such a wonderful example. One of Einstein's many great achievements, and perhaps his greatest, was the General Theory of Relativity which, as I mentioned in my introduction to this book, replaced our 'classical' notions of space and time with something very different – a curved space-time that both explains why things move along the trajectories they do (and not just things like planets, stars and galaxies but also satellites orbiting the Earth, including those that are used in the GPS systems we now use in our cars and on our mobile phones) and is also deformed and curved by those things (as the famous physicist John Archibald Wheeler put it: 'Spacetime tells matter how to move, matter tells spacetime how to curve' J. A. Wheeler, *Geons, Black Holes and Quantum Foam*, New York: W. W. Norton, 1998, p. 235).

Einstein predicted, using his laws of General Relativity, that if light from a star that lay on the other side of the sun from the Earth were observed close to the sun, then it would be seen to 'bend' around the sun, due to the curvature of space-time. Arthur Eddington (at the time, perhaps the world's greatest astronomer) set out to observe the predicted bending of starlight and thus test Einstein's theory, during an eclipse (otherwise he and his team wouldn't have seen anything and might possibly have been blinded!). The rest as they say is history (and fascinating history at that!) but I'll leave you in suspense about the outcome until the next chapter.

It seems that all of these examples can be accommodated by the D-N account: we begin with some laws and some background facts or data and we deduce a prediction, which can then be observed, or not, and thus the laws confirmed, or not (but again, I'll leave that for the next chapter). The basic arrangement seems to be the same as for explanation. However, an obvious difference is that whereas we typically aim to *explain* what has already been observed (e.g. rainbows), we *predict* what we have yet to see (even though astronomers had already observed Neptune before the prediction, they hadn't realized it was a new planet). So, the only difference between explanation and prediction on the D-N account is timing.

This feature associated with predictions will turn out to be important when we come to Chapter 8 on realism. There we will consider an argument to the effect that the prevalence of novel predictions in science – that is, predictions of things that haven't been observed before – should compel us to take seriously the theories used to make those predictions: so seriously in fact, that we should regard those theories as true, otherwise it would have to be a massive coincidence that they could get it right about what has yet to be observed. This is a powerful argument but as we'll see not everyone is convinced.

Coming back to the D-N view, I hope I've indicated what a strong and useful view it is, combining laws and deduction in a very powerful way and bringing together explanation and prediction in one package. Unfortunately, however, it suffers from a major flaw.

The flagpole counter-example

Imagine the following situation: you decide to put up a flagpole in your front yard/schoolyard/university quad … (kids, get permission before you do this!). Perhaps you want to ensure that the shadow of the flagpole doesn't fall across your neighbour's prize-winning flowerbed so you'd like to be able to predict the length of the shadow at certain times of day. Or, more generally, you'd just like to explain why the shadow is as long as it is. The D-N account easily accommodates this scenario: given the laws of geometry, the height of the pole and the angle of the sun, it's a pretty simple matter to deduce the length of the shadow. With that, the length is explained or predicted, depending on when you make the deduction.

So far, so what? Well, using the laws of geometry, the angle of the sun and the length of the shadow, I can also deduce the height of the pole – that's an equally simple matter. And again the D-N view accommodates this straightforwardly, since again we have laws + facts and all we need to do is crank the handle of deduction. But that doesn't seem right! We surely don't *explain*, much less predict, the height of the pole on the basis of the length of its shadow, do we?! After all, the reason why the pole is the height it is has to do with the way it was made, maybe what kind of flag was intended to fly from it, how big and heavy that flag is and so on (we'll leave to one side, for now, examples where a pole or monument is designed to be a certain height precisely in order to cast a shadow of a certain length! Think of sundials for example …). So, the D-N view gets it wrong by including, as explanations, some deductions that we intuitively would not regard as explanations at all. It is too permissive.

Here's another example: Back in the days before we all had weather apps on our phones, many people would have barometers on their walls at home which measured the local air pressure. And since low pressure is generally correlated with low weather fronts, storms and rain, a falling barometer needle was a good indication of rain coming. Now, although we might use the change in position of the barometer needle, together with the laws of meteorology, to predict the rain coming, it seems bizarre to suggest that we *explain* the rain by the movement of the needle (so here the nice package bundling up prediction and explanation comes apart). However, we can certainly deduce that the rain will come or has come from the laws of meteorology plus certain facts pertaining to the material used in the barometer and how it reacts to low pressure etc. And given that deduction, according to the D-N view, this has to count as an explanation. Again, it seems the D-N account has come off the rails.

What has gone wrong? One reason for the mis-match between our intuitions about what should count as an explanation and the D-N account might be that our intuitions are focusing on what *causes* the phenomenon to be explained, and that's what we value in an explanation, whereas the D-N account doesn't pick out or identify the relevant causes at all. So, in the case of the barometer and the coming rain, it's the low pressure associated with the rain that causes the barometer to fall, not the other way around – indeed, it seems utterly implausible that the change in position of the barometer needle could cause the rain to come! Likewise, we take the height of the flagpole to cause – along with the other factors such as the position of the sun – the shadow to have the length that it does. Leaving aside those cases where we put up a flagpole of a certain height precisely because we want the shadow to be a certain length at a given time of day (to annoy our neighbour perhaps), we don't generally take the length of the shadow to be the cause of the height of the pole (but I'll come back to this at the end).

So, one response to the problem with the D-N view is that it misses out the relevant causes and these are what are important in explanation.

Causal accounts

The core idea of these sorts of accounts has to do with what an explanation is: it is not a deduction of a statement describing some phenomenon from laws and initial conditions: rather it involves showing how the phenomenon is a result of a particular causal process or, more generally, how it fits into one or other of the causal patterns that we find in the world. So, to return to our barometer example, the needle of the barometer moves as a result of the process that involves the atmospheric pressure falling, which we describe in terms of a front moving in or a storm passing through. The fall in pressure is the cause and the needle moving is one effect, the rain another. In these kinds of situations, where a cause leads to more than one effect, we talk about the 'common cause' of the effects. So, the low atmospheric pressure is the *common cause* of both the rain and the shift in the barometer needle. The flagpole example is simpler in this respect: here the height of the flagpole (together with the position of the sun of course) causes the shadow to be a certain length and not vice versa (or, usually not vice versa). So the causal view escapes the problem that the D-N view has to face – since causes lead to their effects and not the other way around, we can't run the explanation in both directions, as we can on the D-N account.

We can also understand predictions in these terms: they are just effects that we have yet to observe in the world. So, Einstein's prediction of the bending of starlight around the sun can be understood as the mass of the sun causing the surrounding space-time to become distorted and that distortion causes the starlight to deviate. We explain, on this view, by taking the relevant phenomenon to be an effect and showing how it can be related by some physical mechanism to a cause.

Notice that on this account we don't have to appeal to laws in our explanation. Some philosophers think this is an advantage. So, consider: I come back to my desk at home and notice that my cup of tea has been knocked over. I observe small-dog-sized paw prints in the spilled tea and offer as an explanation that our pet, aka The Small and Fearful Dog, has jumped up and knocked the tea over. Now that may or may not be *true* (perhaps there was a small earthquake that knocked the cup over and also prompted the dog to jump up – another example of a common cause – although given the lack of earthquakes in Leeds, this is unlikely) but the important point is that this seems reasonable as an explanation – the dog jumping up was the cause, the spilled tea the effect.

Of course, if you were to insist that I explain what happened in terms of the relevant laws, I could perhaps do that: I'd have to give the laws of canine behaviour together with the laws of mechanics, as well as the relevant conditions that describe the situation prompting that jumping behaviour etc.; but leaving aside how complex that kind of explanation would be and also the question whether our understanding of the behaviour of dogs can be encapsulated in laws the way our understanding of mechanics can (this is a serious issue in the philosophy of biology for example), in principle at least I could give such an explanation. In that case, the advocates of the D-N view insist, just citing the cause as explanation is not enough – it amounts to little more than a 'sketch' of an explanation.

There are other concerns that the causal account of explanation must face.

One has to do with the question: What makes a process causal? If to give an explanation of some phenomenon is to show how that phenomenon is the effect of some cause or, more generally, arises from some causal process, then we'd better have some way of identifying those kinds of processes, at least in principle. A standard answer to this question is to appeal to 'mark transmission': the idea is that in causal processes a mark can be passed through the process, down the chain of cause and effect from the initial cause to the final effect. So, think of snooker (or pool) balls colliding with one another. We can explain the movement of a particular ball – a red one, say – by appealing to its collision with the cue ball as the cause of its movement. And if the cue ball were marked with some chalk in the right place, we could see that mark transmitted to the red ball via the collision. More generally, we can understand a 'mark' as any change in the structure of something that can persist through the process leading from the cause to the effect that we are trying to explain.

However, an obvious worry to this answer to the question of what makes a process causal is that the mark may be disturbed or lost altogether through some interaction with something else. So, a sudden breeze blowing through the sweaty air of the snooker/pool room might blow the chalk mark off the cue ball before it collides with the red. Of course, an obvious response would be to roll one's eyes and say 'Look, I'm obviously talking about an idealized situation here so let me specify that causal processes are those for which a mark can be transmitted in the absence of any extraneous interventions (like a sudden wind passing through the snooker/pool hall).' That's fine as it goes but the further worry is that, as with all such appeals to idealization, the answer to our original question and the account on which it depends are now so idealized that they don't apply to many, or perhaps even any, physical situations. Even in tightly controlled laboratory conditions there may be extraneous interventions and insisting that we should ignore the ones that are not causally relevant runs the time honoured risk of introducing circularity into our account (explanation has to do with causal processes; causal processes are those that can transmit marks in the absence of relevant interactions; relevant interactions are those that involve causal processes ...).

Furthermore, even if we sort out that issue, the causal account seems to work best in the kinds of situations exemplified by the snooker/pool balls case. Leaving aside the possibility of sudden gusts of wind, these are situations where the relevant causal processes can be relatively easily discerned and both the cause and effect identified more or less straightforwardly. These are the situations typically described within what is sometimes called 'classical' or Newtonian physics. But much of the world – whether described in everyday or scientific terms – is not like that. Much of what we see in the world around us depends on chemical, biological, economic, political and social processes, and identifying the causal processes or even just the causes in those sorts of cases can be a hugely complex business. Indeed, some would argue that when it comes to social phenomena giving an explanation in terms of causes is just not appropriate.

Just consider the debate over the 'cause' of the First World War, for example: many would say that it was the assassination of Archduke Franz Ferdinand and his wife Sophie, some would point to a more general breakdown in the balance of power across Europe, while others would point to particular countries as bearing the blame (see: http://www.bbc.co.uk/news/magazine-26048324). We can all agree that many factors were involved and picking out one as *the* cause, as we can in the flagpole and barometer cases, reduces a complex and nuanced situation to something that is crude, unhelpful and just plain wrong!

Part of the problem here is that unlike in the case of much of physics, we can't 'run' the WWI situation again in order to discern what really was the cause. When it comes to snooker/pool balls, if we have any doubt as to what caused the red ball to move, we can just set them all up again and start the processes in motion once more. The possibility of repeating an observation, or an experiment more especially, is absolutely crucial in science, not least in order to rule out irrelevant or 'fake' causes and to identify what is actually going on. Many apparently important discoveries, some of them much trumpeted in the press and elsewhere, have turned out to be false, or merely artefacts of the experimental set-up (a 'classic' case is that of 'cold fusion', where it was claimed that nuclear fusion – and the associated release of vast amounts of energy – could be achieved at room temperature: http://undsci.berkeley.edu/article/cold_fusion_01). We'll come back to this issue in Chapter 6. The point is that when it comes to many of the events we are interested in from the perspective of history, politics or the social sciences, say, we can't set things up again and start the processes over (or at least we should all hope we can't in cases like WWI).

That's not to say that social scientists don't try to identify the relevant causes of many social phenomena. Correlations between different phenomena can be identified through something called 'systematic comparative case analysis' (if you're really interested see: http://www.compasss.org/about.htm). This involves detailed and, as the name says, systematic comparisons of different kinds of cases involving different factors and variables and has all kinds of applications in social science, economics, criminology etc. but – and here's a slogan to hang on the wall – *correlation is not causation*. So, to give just one of many fun examples, there is a close correlation between the rise in number of users of Facebook (a popular form of social media for those of you reading this five years from now) and the rise in yields on Greek government issued bonds, but that does not mean that Facebook *caused* the Greek debt crisis. Now of course, economists, sociologists, political scientists and so forth have a range of techniques that help them eliminate such spurious correlations and identify what they hope are the causal relations that actually hold (some of these techniques are also used in the physical sciences and are captured in what is known as Mill's Methods, after the great philosopher and political thinker John Stuart Mill: http://philosophy.hku.hk/think/sci/mill.php). The point is that identifying the causes in these sorts of situations is complicated and certainly vastly less straightforward than in the cases of the snooker/pool hall and the comparatively clean and straightforward world of classical physics in general. Hence, outside of these sorts of environments, the causal view may not be the best choice for an account of explanation at all.

Another, and perhaps even more fundamental, concern is that the notions of cause and causality in general have been driven out of modern physics. So, back in Chapter 1, I briefly mentioned quantum physics which, putting it crudely, describes the world as fundamentally random or probabilistic (there are attempts to understand the theory in ways that avoid this kind of description but I won't go into those here). So, think of radioactive decay for example: although we can know how long it will take half the atoms in a sample of radioactive material to decay (that's given by the material's 'half-life') we can't know when a particular atom of that material will decay. Take carbon-14 for example: that has a half-life of 5,730 years, and in a gram of natural carbon we can detect 14 disintegrations of carbon-14 atoms per minute (don't worry – 99 per cent of the earth's carbon is carbon-12 and most of the other 1 per cent is carbon-13 which is not radioactive). That's incredibly useful for archaeologists and historians, for example, since it enables them to date (more or less accurately) all kinds of ancient artefacts by extrapolating back from their current decay rate. This applies to human remains too: the Iceman (not the Fantastic Four superhero) is a frozen body (given the name Ötzi) found in the Italian Alps in 1991, which was dated to 3300–3100 BC on the basis of the carbon-14 that the body still contained.

However, even though this technique has become an incredibly important tool in archaeology, if we take an atom of carbon-14 we cannot say when *that* atom will decay. And ultimately that's because radioactive decay is a quantum process and is fundamentally probabilistic. So we can't predict when it will decay at any given time, nor can we say *why* it decayed when it did. In other words, for these kinds of cases, we cannot give the cause and so can't provide an explanation.

Now, you might be tempted to say, so what?! The quantum world is so far removed from what we encounter in everyday life and much of science that perhaps we shouldn't worry too much if the causal account can't be applied there. That's perhaps a little hasty. Much of the hi-tech environment in which many of us find ourselves these days depends on quantum processes (think of lasers for example). Or take photosynthesis, the process by which plants and other organisms convert light into chemical energy. Recent research suggests that the efficiency of this process can only be explained using quantum physics (take a look at the press release on this discovery and consider what might be involved in that explanation: http://www.ucl.ac.uk/news/news-articles/0114/090114-Quantum-mechanics-explains-efficiency-of-photosynthesis). Hence, it might seem a bit of a radical step to dismiss a whole area of science as falling outside the remit of the causal account of explanation. And given the increasing importance of such processes, the inability to accommodate them might be taken to reflect badly on this account.

Perhaps then we could generalize the causal account to give us a framework that can be applied across a range of different sciences, physical, biological and social. This is what the 'interventionist' approach tries to do.

Manipulating and intervening

This focuses on the apparently plausible idea that causal processes may be manipulated, controlled and exploited in general. So the core idea is that something is the cause of a given phenomenon if manipulating that something leads to a change in that phenomenon. In other words, varying the cause leads to a variation in the effect.

Suppose I notice that my prized 'Yellow Hammer' dahlias are all wilting and looking sickly. One way of establishing the explanation of this dire situation is to manipulate the circumstances under which the plants are kept. So, I might stop watering them so frequently and lo and behold, they perk up and start looking healthy. I conclude that the cause and hence explanation of their wilted appearance was over-watering. We can more or less straightforwardly extend this kind of account to more complex and significant issues, from the manipulation of genes and other biological entities to social manipulation, which of course can be hugely controversial.

For example, you might be interested – for perfectly good reasons I should emphasize – in the explanation of mood changes among people who heavily use certain social networks such as (again)

Facebook. One way of discovering the cause of this apparent phenomenon would be to manipulate the material posted on such a network by the friends of certain (randomly selected) users by, first, reducing the positive content of those posts and then, reducing their negative content. It turns out that if you do that, you find an increase in negative personal status updates in the first case and an increase in positive status updates in the second. Hence, it can be concluded that changes in mood of the users of such networks may be explained by the emotions expressed by their friends (http://www.wired.com/2014/06/everything-you-need-to-know-about-facebooks-manipulative-experiment/).

However, there is an obvious problem with this kind of manipu-lationist account (and not just that we might have ethical qualms, as with the above example): many phenomena that we would like to explain in science are not open to manipulation. The causes and thus explanations for the destruction of Pompeii and the extinction of the dinosaurs are the eruption of the volcano Vesuvius and meteor impact, respectively (although for a while volcanoes were also thought to be responsible for the dinosaur extinction; we'll come back to that in Chapter 8). There is simply no way those events are now or ever were open to manipulation (at least not by us). And it's not so hard to think up other cases (go on, it's easy!).

One way to respond to this concern is to generalize the idea still further and introduce the notion of an *intervention*: something causes, and therefore explains the occurrence of, a particular phenomenon if, by intervening on that something, some (relevant) feature of that phenomenon would change.

So, for example it was discovered that around 20 per cent of women who have breast cancer also have an abnormal amount of a certain protein called HER2. It was then noticed that the growth of tumour cells in such cases appeared to be suppressed by a certain kind of antibody, called trastuzumab or the easier-to-pronounce name of Herceptin. To see if this was the cause of the cancer suppression, researchers set up a trial: they divided 450 women with this form of breast cancer into two groups and gave one group Herceptin and the other a placebo in a 'double blind' test (so neither the patients nor the researchers knew who was getting the drug and who the placebo, in order to help eliminate possible bias). A statistically significant reduction in the growth of the tumours was identified and over 400,000 women around the world with this kind of breast cancer have now been treated with this drug.

From the perspective of the interventionist approach to explanation, the researchers intervened in the situation and showed that the antibody caused the suppression of the growth of the tumours. If they had not intervened by giving the relevant group of women the Herceptin, their tumours would (most likely) not have been suppressed, as happened with the group given the placebo. Hence, the explanation for the effect is the administration of the drug. Likewise, going back to the earlier example of my spilled cup of tea, I could have intervened to stop The Small and Fearful Dog from jumping up and the cup would not have been knocked over.

Of course, you might worry that this hasn't progressed very much from the idea of manipulability above – after all, how could we have intervened in the case of the destruction of Pompeii or the extinction of the dinosaurs?! However, the notion of an intervention is intended to be more general than manipulability and not human specific. So, something – presumably something geological – could have intervened and blocked or diverted the explosion of Vesuvius and something – another meteor perhaps – could have diverted the meteor that killed the dinosaurs. What this approach does is encourage us to focus on 'what would have happened if things had been different' sorts of questions and because it doesn't invoke laws, whether of the natural or social worlds, or assume any particular causal mechanisms, it seems to be applicable across a whole range of different situations and sciences.

Indeed, this approach has attracted a lot of attention in the philosophy of science in recent years. However, it still doesn't accommodate the example of radioactive decay since there is nothing we can do and more importantly nothing that can intervene to make a particular atom of carbon-14 decay any earlier or later. Although the interventionist account deploys a notion of causation that is much more general and encompassing than other similar accounts, it's still at heart a *causal* approach and that just won't work when it comes to modern physics.

But do not despair! Here's an approach that does.

Explanation as unification

Another notable feature of scientific advances is that they often bring together under some unified framework apparently quite different phenomena. We have already encountered some well-known examples of this in Chapter 3, when we considered discovery: Newton's theory of gravitation which unified the falling of objects on earth with the orbit of the planets around the sun and Maxwell's theory of electromagnetism, which brought together under one conceptual scheme electricity and magnetism. To these we might add the unifications achieved in biology by the theory of evolution and genetics. And by bringing these phenomena together under one unified theoretical framework, we explain them, or so it is claimed. Hence, we explain why objects on earth (apples, people, Small and Fearful Dogs) fall the way they do and why the planets orbit the sun they way they do in terms of a common conceptual scheme, involving gravity. On this account, radioactive decay is explained in terms of its falling under the best and most unificatory conceptual scheme we currently have, namely quantum theory.

Furthermore, the more phenomena we can bring together this way, the more powerful our explanations will be. Eventually if we were able to come up with the Grand Unified Theory of everything, including quantum processes, gravity and also social, economic and political phenomena – something that many philosophers have doubts we will ever be able to achieve – then we would have the Ultimate and Most Powerful explanation of all (mwahahahaha!).

Notice two things however. First, the notion of a phenomenon 'falling under' a conceptual scheme is doing a lot of work in this approach. Can we make this notion more precise? Well, yes we can – via deduction. Both the fall of apples and the orbit of Venus can be explained by deducing statements about them from Newton's unifying theory of gravitation. What is this theory? Well, at its core it has the famous law of gravitation that states that gravitational attraction is proportional to the product of the masses of the objects and inversely proportional to their distance apart. But now it seems we're back to a laws plus deduction approach – been there, done that!

And that takes us to the second worry. Let's go back to our old friend, the flagpole. How does this approach accommodate that example? We could presumably construct some kind of unified account of how objects cast their shadows under different conditions but it's unclear how that would help us explain why the length of *this* flagpole's shadow is what it is. More generally, and again like the D-N view, this approach does not seem to capture the kinds of causal relevance that we typically take to be significant, or at least it does not do so straightforwardly. However, proponents of the unificationist approach insist that their opponents have got the relationship between causation and explanation back to front: typically explanations form answers to 'why?' questions and do so using 'because'. Why did the apple fall to the ground? Because gravity caused it to ... Those advocating the causal approach point to this and say 'Aha! See, that shows that explanation must be based on causation.' The unificationists, on the other hand, insist that the causal folk have got it the wrong way round and the 'because' they point to as indicative of causation is just a subsidiary feature of explanation, which is more general and unificatory. According to this sort of view, appeal to causation is just a kind of holdover from our ancestors' crude attempts to explain all sorts of phenomena (in terms of gods, spirits, occult forces and so on ...) but now we can liberate ourselves from such a narrow focus!

Still, it's hard to shake the sense that causation has something to do with explaining things, at least in many cases. Perhaps, then, we should give up the idea that one account of explanation can fit all cases.

Pluralism and pragmatism

On the basis of even this quite sketchy survey of different possible accounts of explanation, some might conclude that it's a case of 'horses for courses', with different accounts suitable for different areas of science, or even for different domains within the same area. So, perhaps the causal account is more suitable for big chunks of 'classical' or Newtonian physics, where we might think of the world in terms of snooker-ball-like atoms bouncing off one another and a chain of cause and effect running from the beginning of such a process to the end. But then when we come to think of explanations on a grander scale, like those that result from the bringing together of electricity and magnetism under one conceptual framework, or planets and apples, then it may be that the unificationist account does a better job of capturing the kinds of explanations that are used in those situations.

Spreading our wings further, many explanations in the life and social sciences refer to the function of certain entities within some larger system. I don't have the space to go into these sorts of explanations here, but the general idea is that we explain why certain features or elements are present in a system by referring to the function of these features or elements. So, here's a well-used example: Why do we have hearts? Because their function is to pump blood and we need to have our blood pumped around the body in order to stay alive. Of course, other elements or features could perform the same function, so you might feel this kind of explanation is a bit inadequate or, at least, incomplete.

Some philosophers believe that this sort of account can be strengthened by combining it with aspects of the causal view and argue that in many cases – particularly in the life sciences – the functions we are interested in can be explained via certain kinds of *mechanism*. So the central idea is that the relevant phenomena are explained by describing the mechanisms involved in producing them. Of course, that requires us to say something about what a 'mechanism' is and there is on-going debate about this, but broadly speaking it consists of a set of entities plus the activities they engage in. As an example, consider the circulatory system of the human body: there we have various entities such as the heart, the lungs, the blood vessels themselves, engaged in the activity of delivering blood – and hence oxygen, nutrients etc. – to the various parts of the body, thus forming a complex mechanism, and through the activities and interactions of these various entities specific functional behaviour results.

Whether this is really different, at its core, from the causal account remains debatable, as is the set of issues around what constitutes a mechanism, but this philosophical analysis is currently being applied to a range of different domains, including psychology and neuroscience, as well as biology (where, as I mentioned above, we typically don't find the kinds of laws we have in physics so the D-N view can't get much of a grip).

Even if you accept that these different accounts are not really competitors, since they apply in different scientific fields, or domains, or to different areas within a domain (so in that sense you would be a *pluralist* about explanation), you might feel that within a given field, domain or area, whatever kind of explanation applies to that field/domain/area *gets it right*, in the sense that it gives us 'the truth' about why a particular phenomenon occurs where or how or in the form it does. Indeed, you might feel that that's partly – maybe wholly – the aim of giving an explanation in science in the first place. After all, what's the point of giving an explanation if not to tell us what's 'really going on'?!

We'll come back to these issues to do with truth and reality in Chapters 8 and 9 when we discuss the realist–anti-realist debate, but here I just want to note that not every scientist and certainly not every philosopher of science thinks that the aim of explanation is to get at the truth, or at least come close to it. Some adopt a form of 'pragmatic' view which holds that even within a particular field, or domain or area or whatever of science, there is no one explanation leading to 'the truth'. Instead there are typically multiple explanations and we can choose which one to give prominence to, based on various contextual or external factors. Some then argue that there is not, and never was, any truth to be had, or that we can gain access to – and we'll come back to that sort of view in Chapter 9 – whereas others suggest that these different explanations all offer different perspectives on 'the truth' and may or may not be integrated depending on the circumstances.

Consider the barometer example again and the question 'Why has the barometer needle dropped?' According to certain pragmatic views of explanation, we need to establish the relevant context by considering, first, whether the question is asking about a particular barometer needle and why it dropped rather than rose or whether it is asking why this one dropped while other ones didn't. Second, we need to establish the relevance of our possible answers by considering what information is being sought: if it's causal information we want, for example, then given the first type of answer above we might give the standard answer in terms of an approaching low-pressure system. But perhaps we're seeking information about the function of the barometer in the context of a bad weather early warning system. Third, we can then evaluate the answers and hence explanations given, again in the particular context. On this kind of account, then, explanation is not just a matter of connecting laws or causes or functions and phenomena; rather, it is a three-way relationship involving the *context* as well.

The point is, since explanation is dependent on context in the ways just indicated, there is no 'right' explanation for any given phenomenon – it all depends on what the question is understood to be asking, what kind of information we want and on our evaluation of the answers we get. Let's go back to the flagpole case and the side remark I made about sundials and let's consider a (literally) concrete example: on Guernsey, one of the Channel Islands between the UK and France, there is a monument to commemorate the liberation of the islands from occupation during World War II. The monument is set up so that each year on the 9th May – the day of liberation – the tip of its shadow reaches a curved white bench on which the crucial events of that day are recorded, including Winston Churchill's statement announcing the islands' liberation (see: http://www.astronomy.org.gg/liberation.htm). In this particular case, the height of the monument *is* explained by the length of the shadow, as it was precisely designed to be that height in order for the shadow to be long enough to move along that bench. And of course, it only does so on that particular day, in that particular location. So, whether we take the height of the monument or flagpole to explain the length of the shadow, or vice versa, depends on the relevant context, and, given that point, there can be no 'right' explanation.

An obvious worry about this sort of account is that its usefulness depends on what determines or delineates the relevant context. It shouldn't be a case of 'anything goes' because then we wouldn't be able to distinguish explanations from arguments or descriptions that aren't explanatory, even given the pragmatic factors. At least when it comes to scientific explanations we can insist that the relevant context must have to do with the appropriate scientific background knowledge. However, it then seems that this sort of account simply comes down to saying that a scientific explanation involves nothing more than showing that certain scientific claims are relevant to understanding the given phenomenon, where that relevance is established using the appropriate background information. *That* seems a pretty general and therefore pretty weak sort of claim; but perhaps that's the best we can get!

There is of course much more to say, but let's move on now and consider how the theories, hypotheses and models that are used in explanations are tested and confirmed or falsified by the evidence.

5

Justification

Introduction

Having used your hypothesis, model or theory to *explain* some interesting phenomenon – according to one or other of the accounts of explanation just presented – the issue now is whether that phenomenon can or should then be cited as *support* for your hypothesis or whatever. Of course, many of the phenomena that science explains can be observed quite easily – just by stepping outside on a rainy day in the case of rainbows. But much of what science does involves laboratory work, in an environment where phenomena are effectively created, so they can be studied and manipulated and carefully controlled. The experiments that feature in this lab work typically generate data (sometimes huge amounts of it, that can only be analysed by massive computers, and from experiments that require teams of hundreds of people). Here, then, is our fundamental question: What is the impact of experimental data on theories? We will look at two answers. The first states that the role of data is to *verify* theories; the second insists that, on the contrary, that role is to *falsify* theories. Let's consider these two answers, before moving beyond them.

Verifiability is what it's all about!

The first answer to our question was most famously propounded by a motley group of philosophers, scientists, economists and others who came to be known as the 'Logical Positivists'. They were called this, first, because they were seen as part of a line of commentary on science that emphasized scientific knowledge as the highest, in some sense, or most authentic form of knowledge, obtained via the *positive* support given to theories by observations through the scientific method; and second, because they deployed the full resources of logic, and in particular the formalization of logic made available by the likes of David Hilbert, Bertrand Russell and Alfred North Whitehead, in the early part of the twentieth century, to both analyse and represent this form of knowledge.

There is a crucial question, which will help us get a handle on what the logical positivists were all about: What is the difference between metaphysics and physics? (or more generally between philosophy and science?). This was a question that fundamentally bothered the logical positivists at the turn of the last century. On the one hand, science seemed to be making huge advances in explaining the natural world; on the other hand, philosophers were developing ever more elaborate metaphysical schemes which seemed unfettered by the kinds of constraints science was seen as working under. Here's one way of understanding the difference between metaphysics and physics; consider the difference between the following kinds of questions:

Metaphysical 'What is the nature of Being?' As far as the positivists were concerned, these sorts of questions had no definite answers, with philosophers unable to agree not only on the answer but on the very grounds for determining what constitutes an adequate answer.

Physical 'Does light bend around the sun?' This kind of question not only seemed to have a definite answer but scientists appeared to agree on the criterion for adequacy – namely *verifiability*. Here the influence of Einstein's work was crucial.

This is how Rudolf Carnap, one of the most famous and significant of the logical positivists, put it:

During a quiet period at the Western Front in 1917 I read many books in various fields, e.g., about the world situation and the great questions of politics, problems of *Weltanschauung*, poetry, but also science and philosophy. At that time I became acquainted with Einstein's theory of relativity, and was strongly impressed and enthusiastic about the magnificent simplicity and great explanatory power of the basic principles.

(R. Carnap, 'Intellectual Autobiography', in P. A. Schilpp, *The Philosophy of Rudolf Carnap*, LaSalle, IL: Open Court, 1963, p. 10)

This is a striking image: hunkered down in the German trenches, amidst the mud and the horrors of war, Carnap read about Einstein's General Theory of Relativity, which, as I mentioned in Chapter 4, fundamentally changed our ideas of space and time, suggesting that matter could curve the space-time around it, and curved space-time led to changes in the trajectories of bodies. Such bodies could be material and immaterial, as in the case of photons, from which light is composed. It was this effect that underpinned the most striking verification of Einstein's theory in the immediate years after World War I. General Relativity predicted that a massive object, such as the sun, would distort space-time and deflect a light ray, from distant star, for example. In 1919, Eddington, a well-known British astronomer, and his team observed just such a deflection and almost exactly to the extent predicted by Einstein's theory (the BBC produced a somewhat over-dramatized TV show about this called, creatively enough, 'Einstein and Eddington', which you can buy on DVD). Suddenly Einstein became a household name, perhaps because of the radical change in our view of space and time proposed by his theory, or perhaps because in the post-war world, the public were simply taken with the idea of a Swiss-German physicist having his theory confirmed by a British astronomer (and a Quaker to boot).

More importantly, as far as we are concerned, it had a tremendous impact on the positivists and supported their crucial idea that what distinguishes scientific theories from metaphysics, poetry etc. is their observational *verifiability*. It is this that demarcates science from other human activities. Not surprisingly, perhaps, we find similar sentiments expressed by scientists themselves; here is a report on the attitude of August von Hofmann, a famous chemist in the nineteenth century, who discovered several organic dyes, and was the first to introduce the term 'valence' (to describe the capacity of atoms to combine) and to use molecular models in his lectures:

> Prof. A. Senier relates that von Hofmann used to say to him and other research students working in the Berlin laboratory, "I will listen to any suggested hypothesis, but on one condition – that you show me a method by which it can be tested." Without such a condition, the creations of a disordered mind would be as worthy of consideration as the speculations of a scientific genius; and fertile ideas would be sought not in a laboratory but in an asylum. A hypothesis ought, therefore, to be capable of being verified, even though the means may not be available of applying a crucial test to it at the time.
>
> (R. Gregory, *Discovery: Or The Spirit and Service of Science,* London: Macmillan, 1923, p. 162)

Verifiability as demarcation

According to the logical positivists, then, to be scientific a hypothesis must be capable of being verified: that is, it must be *verifiable*. It is verifiability that demarcates science from non-science. This seems a nice and clear way of sorting the scientific wheat from the non-scientific chaff. Indeed, the positivists went even further and insisted that for a statement to be *meaningful* it had to be *verifiable*: that is, there had to be at least the possibility that it could be verified, in principle.

Here's how another well-known positivist put it:

> The most celebrated example of this [meaning in terms of verifi- ability], which will forever remain notable, is Einstein's analysis of the concept of time, which consists in nothing else whatever but a statement of the *meaning* of our assertions about the simultaneity of spatially separated events. Einstein told the physicists (and philosophers): you must first say what you *mean* by simultaneity, and this you can only do by showing how the statement 'two events are simultaneous' is verified. But in so doing you have then also established the meaning fully and *without remainder*. What is

true of the simultaneity concept holds good of every other; every statement has a meaning only insofar as it can be verified; it only *signifies* what is verified and absolutely *nothing* beyond this.

(M. Schlick, 'Positivism and Realism', in R. Boyd et al. (eds), *The Philosophy of Science*, Cambridge, MA: MIT Press, 1991, p. 42)

This is a strong criterion: too strong, perhaps. Consider the statement, 'For a statement to be meaningful, it has to be verifiable.' Is this meaningful? Intuitively it would seem to be (even if you don't agree with it) but is it verifiable? It would seem not; after all, what could possibly verify it? But then, by their own criterion, the central tenet of the logical positivists' view is meaningless!

This might seem a bit of a 'philosophical' objection to what seemed to be an intuitively clear feature of scientific practice, but there are others, as we shall see.

From verification to confirmation

We recall the hypothetico-deductive picture: we come up with a hypothesis. (How? Who knows? Who cares? At least that's what Popper would say.) Then we deduce a possible observation of some phenomenon, giving us a prediction (at least if the phenomenon has yet to be observed). Now, according to the verifiability approach, if this phenomenon is actually observed, we have verification of the theory/hypothesis. Does this verification lead to the truth? No; one verification does not the truth make – the very next prediction might not be observed and the hypothesis would be regarded as false. A more plausible view would be to say that the greater the number and variety of verifications, the greater the support for the theory and the higher the *probability* of being true (recall the inductive picture).

What this suggests, however, is that a hypothesis can never be completely verified and so the verifiability view needs to be modified. With this in mind, the logical positivists began to shift their emphasis from the verification of a hypothesis to its confirmation. Here's Carnap again:

Hypotheses about the unobserved events of the physical world can never be completely verified by observational evidence. Therefore I suggested that we should abandon the concept of verification and say instead that the hypothesis is more or less confirmed or disconfirmed by the evidence. At that time [1936] I left the question open whether it would be possible to define a quantitative measure of confirmation. Later I introduced the quantitative concept of degree of confirmation or logical probability. I proposed to speak of confirmability instead of verifiability. A sentence is regarded as confirmable if observation sentences can contribute either positively or negatively to its confirmation.

(R. Carnap, 'Intellectual Autobiography', in P. A. Schilpp, *The Philosophy of Rudolf Carnap*, LaSalle, IL: Open Court, 1963, p. 59)

So, the more evidence we have, the more the hypothesis is confirmed. This seems a plausible view and it accords with the common-sense perception that science is built on the 'facts'. However, it faces a number of problems – some of them are specific to this approach and some have to do with this common-sense perception, as we shall see.

Problems

First, consider the following question: Are statements verified in isolation? The above discussion appears to have assumed that they are, in that we take a hypothesis and then consider how the evidence verifies or confirms it. But it's not hard to see that this is far too simplistic. Consider again Einstein's hypothesis that the curvature of space-time around the sun leads to the bending of starlight. What was involved in the testing of that hypothesis? Various assumptions had to be made about the orbit of the earth around the sun, about the movement of the earth and the sun relative to the stars and so on. Various bits and pieces of apparatus had to be brought together in order to make the observations, and in order to understand these, various other hypotheses had to be understood too. In other words, the experimental test of a theory requires various 'auxiliary hypotheses' in order to 'hook' the theory or hypothesis up with the evidence. So, what is actually being verified or confirmed? Clearly not the original hypothesis on its own: rather it is the whole network of hypotheses, the original one plus the auxiliaries that are related to the evidence. And so, verifiability cannot be a criterion of meaningfulness for individual statements but only for a whole network. This is known as the 'Duhem-Quine problem' and it's a problem because first of all, we can no longer talk of individual statements being meaningful or not, but only whole networks of inter-related hypotheses, And secondly, it's a problem because, shifting from verification to confirmation, the confirmation of individual hypotheses no longer makes any sense, but only the confirmation of these inter-related networks. Now this view seems less plausible.

Second, consider this question: How many observations are needed to verify/confirm a hypothesis by a given amount? In some cases, a number of observations are required before a given hypothesis is deemed to be sufficiently confirmed for scientists to accept it. In others, we need only one. Consider the hypothesis that fire burns, for example! More seriously, Einstein's hypothesis regarding the bending of light was taken to be dramatically confirmed by Eddington's observations. Of course, that hypothesis was only part of the grand Theory of General Relativity but nevertheless the confirmation was taken to be stunning and, if not conclusive, certainly not as requiring a whole slew of further observations. And even in cases where further observations are made, and are seen to be needed, further questions arise, such as when does a further observation count as a new one and not merely a repetition of one already made? It seems plausible to say that the more different observations are made, the more a hypothesis is confirmed and that merely repeating the same observation should not be understood as further increasing the confirmation of the hypothesis. But then how, precisely, are 'new' observations to be distinguished from mere repetitions?

Of course, scientists themselves will have a lot to say about this sort of question, and often engage in heated debates about the significance of certain observations, but the issue for us is whether the distinction between 'new' observations and repetitions can be easily accommodated in our account of confirmation. It was precisely concerns such as this, concerns that had to do with the issue of how various features of actual scientific practice can be captured by our philosophical accounts, which led to the ultimate decline of the positivist view. And, in particular, as we shall see, the implicit reliance on observations as some kind of bedrock of scientific objectivity proved to be much more problematic than originally anticipated.

But before we get to that point, let us consider a related view, one which has proven to be more successful than positivism, among scientists themselves at least.

No, no, it's falsifiability!

This alternative view was articulated around the same time as logical positivism and was similarly strongly influenced by the dramatic scientific achievements of Einstein. This view, known as 'falsificationism', for reasons that will shortly become clear, was single-handedly developed by Karl Popper, who started off aiming to become a primary school teacher, got his PhD in philosophy and was friendly with some of the logical positivists. However, the impact of three 'theories' – the Marxist theory of history, Freudian psycho-analysis and Adlerian psychology – and the comparison with Einstein's theory of relativity led him to reject verificationism entirely. What bothered him about all three theories is that they seemed to be supported by what he called 'an incessant stream of confirmations': every political event reported in the news, even the way it was reported, was taken by the Marxists of Popper's time as supporting their view that all political, social, cultural and, yes, scientific structures were determined by the mode of economic production; and the Freudian psycho-analysts seemed to constantly emphasize how Sigmund Freud's theories of unconscious repression or the role of the Oedipus complex (in which the child becomes sexually fixated on the mother) in the formation of neuroses were verified by their clinical observations. As for Alfred Adler, Popper worked with him briefly, helping socially deprived children, and reported describing a case which did not seem to him, Popper, to fit Adler's theory, but which the psychologist apparently had no difficulty explaining in terms of his theory of the role of the inferiority complex. This did not impress Popper at all, as it seemed to him that all it confirmed was that a particular case could be interpreted in the light of a theory.

Popper invites us to consider two examples of human behaviour in order to illustrate what he's driving at: the first is that of a man who pushes a small child into a lake with the intention of drowning it; the second is that of a man who sacrifices his life in an attempt to save the child. Each case can be easily explained by either Freudian or Adlerian psychology. According to Freud, the first man suffered from repression, perhaps of some aspect of the Oedipus complex, in which his prohibited desire, as a child himself, for his mother, effectively manifests itself by a roundabout path, leading to an act of violence; whereas the second man's act can be explained by his having achieved sublimation, by which the unwanted impulses are transformed into something less harmful, indeed, are channelled into an act of heroism. As far as Adler is concerned, the first man's act can be explained by his intense feelings of inferiority, leading to an equally intense need to prove himself by some means, such as committing a horrendous crime; the second man's heroism can be explained in the same terms, his inferiority complex compelling him to prove himself by attempting a daring rescue. The example can be generalized: there is no aspect of human behaviour that cannot be interpreted by either theory. And it is this feature, that the facts could always be fitted to the theory, that was presented as the real strength of these theories. However, Popper insisted, this is not a strength at all, but a disabling weakness.

Compare these examples to that of Einstein's theory and the prediction that starlight will be bent around the sun. As with the positivists, this was something that made a big impression on Popper. And what impressed him the most was how risky Einstein's prediction was: Eddington might have sailed off, made his observations and revealed that starlight was not bent around the sun, that the prediction was, in fact, false. And in that case, Popper insisted, the hypothesis and indeed the whole theory of General Relativity would have been *falsified* and *this* is what renders Einstein's theory scientific and both Freudian and Adlerian psychology mere pseudo-science.

So, let us compare and contrast:

Adlerian psychology No matter what happens, the theory has an explanation. And so, there is no possibility of it ever being wrong: any and all phenomena can be explained and encompassed by it. But if it is consistent with *any* kind of human behaviour, then it tells us *nothing* about human behaviour.

Einstein's General Theory of Relativity This theory makes definite predictions that could turn out to be wrong and thus allow for the possibility of General Relativity being false; in other words, the theory is *falsifiable*. And it is this characteristic that distinguishes science from metaphysics, and 'proper' science from fake or 'pseudo-' science. Let us look at this idea more closely.

Falsifiability as demarcation

According to this view, to be scientific a hypothesis must be capable of being falsified – that is, it must be *falsifiable* – in the sense that it makes definite predictions that could turn out to be wrong. Hence, it is falsifiability, not verifiability, that demarcates science from non-science. Consider the following examples:

Falsifiable	Unfalsifiable
It always rains on Mondays	It is either raining or not raining
All swans are white	All bachelors are unmarried
The gravitational attraction between two bodies is proportional to the product of their masses and inversely proportional to the square of the distance between them	Love breakups are possible, some dishonesty in another's attitude [from a horoscope]

The claim that it always rains on Mondays, although apparently plausible when uttered on a grey March day in Leeds, is easily checked by waiting for Monday and making the relevant observation, whereas 'It is either raining or not raining' is always true, whatever the conditions outside are like (it is what logicians call a 'tautology'). Similarly, the insistence that 'All swans are white' can be (less easily, if you live in the UK) shown to be false by heading off to Australia (or perhaps a decent zoo, if you like those sorts of places) and observing the famed black swans of Queensland. However, there is no possibility of doing anything similar with 'All bachelors are unmarried', since that can never be false: it is true by definition of the word 'bachelor'.

The third example is more contentious. 'The gravitational attraction between two bodies is proportional to the product of their masses ...' is a (partial) statement of Newton's Law of Universal Gravitation and we can again easily imagine how it might have been false – it might have been, for example, that the gravitational attraction is not inversely proportional to the square of the distance between the masses, but inversely proportional to the square plus a little bit more, 2.05, say. Or, instead of being entirely attractive and proportional to the product of the masses, there may have been a repulsive component; and indeed, something similar to this was suggested some years ago, although found not to be true. Now, the problem with 'Love breakups are possible' by comparison, is not that it is true by logic or by definition (unless you have a very cynical view of love) but that it is so vague and unspecific that it is hard to see how it might be shown to be false. Here's another example, taken from 'my' horoscope in a tabloid newspaper: 'Money issues will be important today.' Well, when is money not important! I might be worried about my pay, about the amount I just forked out for my new iPhone or, more boringly, about my mortgage. Again, it's hard to see under what circumstances such a statement might be falsified, and this is the problem with astrology, namely that its claims are often too non-specific to be falsifiable, and hence they rule nothing out and so tell us nothing (and when they are specific, they're just false, like the prediction by a famous tabloid astrologer that a new object discovered in the solar system beyond Pluto would be classified as a planet, something that you might think would bother astrologers; it wasn't). For that reason, on this view, astrology would not be counted as scientific.

Let's put the philosophical cat among the psychological pigeons. Into which category – falsifiable or unfalsifiable – does the following fit?

At any rate, one can give a formula for the formation of the ultimate character from the constituent character-traits: the permanent character-traits are either unchanged perpetuations of the original impulses, sublimations of them, or reaction-formations against them.

(S. Freud, *Character and Culture,* New York: Collier, 1963, p. 33)

What Freud is talking about is the formation of our adult personalities – to put it crudely – on the basis of certain infant character traits. And what he seems to be saying here is that our adult characters are either straightforward continuations of our juvenile ones, or sublimated forms of them, or based on reactions against them. So, I am what I am, because I was what I was as a child; or because I have sublimated what I was as a child or because I am reacting against what I was as a child. Has he left anything out? If you were to look at the personality of the person sitting next to you and you had access to some record of what they were like as a kid, is there any possibility of Freud's hypothesis being shown to be false? By ruling nothing out he appears to have rendered his hypothesis unfalsifiable and hence, on this view, Freudian psychology, like astrology, does not count as scientific.

Now that's a contentious claim and psychologists can respond either by reforming Freudian theory so as to make it falsifiable or they can question the Popperian view itself and reject falsifiability as the criterion for demarcating science from pseudo-science (or they can drop Freud altogether and come up with a better theory!). As we shall see, there are grounds for concluding that the Popperian view is not as straightforward as it might seem and that falsifiability is not the best way of characterizing the scientific enterprise. Before we get to that point, however, let us linger a little longer and appreciate some of the virtues of the 'falsificationist' position.

It's worth noting, for example, that it ultimately rests on a simple (logical) point: you can never prove a theory to be true by accumulating more and more positive observations (that is, by induction), since no matter how many white swans you have observed, there may always be more swans out there you haven't observed, and some of those may be black. However, you can prove a theory false, by observing just one black swan, for example. Here's how Popper himself put it:

Scientific theories can never be 'justified', or verified. But in spite of this, a hypothesis A can under certain circumstances achieve more than a hypothesis B – perhaps because B is contradicted by certain results of observations, and therefore 'falsified' by them, whereas A is not falsified; or perhaps because a greater number of predictions can be derived with the help of A than with the help of B. The best we can say of a hypothesis is that up to now it has been able to show its worth, and that it has been more successful than other hypotheses although, in principle, it can never be justified, verified, or even shown to be probable. This appraisal of the hypothesis relies solely upon deductive consequences (predictions) which may be drawn from the hypothesis. There is no need even to mention induction.

(K. R. Popper, 'On the So-Called "Logic of Induction" and the "Probability of Hypotheses"', *Erkenntnis* 5 (1935): 170ff.; reproduced in *The Logic of Scientific Discovery*, London: Hutchinson, 1959, p. 315)

Note that the best we can say about a hypothesis, according to Popper, is not that it is true, but only that it has 'shown its worth' up until now. Hypotheses and theories can only ever be provisionally accepted, since the possibility of falsification lies just around the corner. This gives rise to a fairly straightforward view of how science works, and just as with positivism it is one that has proved attractive to many scientists themselves. Here is Oppenheimer, a physicist and so-called 'Father of the Atom Bomb', talking about Freud:

... one of the features which must arouse our suspicion of the dogmas some of Freud's followers have built up on the initial brilliant works of Freud is the tendency toward a self-sealing system, a system, that is, which has a way of automatically discounting evidence which might bear adversely on the doctrine. The whole point of science is to do just the opposite: to invite the detection of error and to welcome it. Some of you may think that in another field a comparable system has been developed by the recent followers of Marx.

(R. Oppenheimer, 'Physics in the Contemporary World', lecture given at MIT, 1947)

How science works (according to the falsificationist)

Let's recall Popper's view of scientific discovery: it's not the job of the philosopher of science to worry about how scientists come up with their theories and hypotheses. It could be through dreams or drugs but, however they are arrived at, it's not a rational process, unlike the 'justification' phase, when the hypothesis is confronted with the evidence. This is what we should be looking at, according to Popper, and it is in that confrontation that the rationality of science lies. And the crucial aspect of the relationship between the hypothesis and the evidence has to do with its potential for being falsified.

So, here's how science works, according to the falsificationist: we start with a hypothesis, arrived at by whatever means. From that we obtain, by logical deduction, a prediction about some empirical phenomenon. If the prediction is incorrect, the hypothesis is falsified and we come up with another. If it's borne out, we take the hypothesis to have 'shown its worth' for now, and test it again, until it too breaks and we come up with another, better one. In this way, science progresses. Here, again, is what Popper says:

> A scientist, whether theorist or experimenter, puts forward state-ments, or systems of statements, and tests them step by step. In the field of the empirical sciences, more particularly, he constructs hypotheses, or systems of theories, and tests them against experience by observation and experiment.... I suggest that it is the task of the logic of scientific discovery, or the logic of knowledge, to give a logical analysis of this procedure; that is to analyse the method of the empirical sciences.
>
> (K. R. Popper, *The Logic of Scientific Discovery*, London: Hutchinson, 1959 [1934], p. 27)

The best hypotheses, on this view, are the ones that are highly falsi-fiable, because they are not vague but specific and make precise predictions and hence tell us more about the world. These are the hypotheses, like Einstein's, that 'stick their necks out', and make bold conjectures. Here, then, is the full falsificationist picture.

Conjectures and refutations

We start off faced with a scientific problem, such as some phenomenon that needs explaining, and in order to solve the problem and explain the phenomenon we come up with a bold conjecture. From that, we deduce some observational consequence relating to the phenomenon, which forms the basis of an experimental test. If the hypothesis passes the test, the conjecture is taken to be corroborated – neither confirmed nor taken to be true, merely corroborated and accepted as the best we have for now. Furthermore, it is accepted only provisionally as we devise more stringent tests. Once the conjecture fails such an experimental test, we take the conjecture to be falsified and so we have to come up with a new one. But in falsifying our bold conjecture and devising a new one we learn about the world, about what doesn't work, and on that basis we make progress.

Darwinian view of science

This can be thought of as a kind of Darwinian view of science, in the sense that hypotheses are thrown to the wolves of experience and only the fittest survive. Here's Popper, again:

> I can therefore gladly admit that falsificationists like myself much prefer an attempt to solve an interesting problem by a bold conjecture, even (and especially) if it soon turns out to be false, to any recital of a sequence of irrelevant truisms. We prefer this because we believe that this is the way in which we can learn from our mistakes; and that in finding that our conjecture was false we shall have learnt much about the truth, and shall have got nearer to the truth.
>
> (K. R. Popper, *Conjectures and Refutations*, London: Routledge and Kegan Paul, 1969, p. 231)

As I mentioned above, many scientists appear to favour this sort of account. One of the falsificationist view's most prominent supporters was Sir Peter Medawar, who was awarded the Nobel Prize in Physiology or Medicine in 1960 for the discovery of 'acquired immunological tolerance', which had a huge impact on tissue grafting and organ transplants. He stated that Popper's classic work, *The Logic of Scientific Discovery*, was '[o]ne of the most important documents of the twentieth century' and wrote,

> The process by which we come to formulate a hypothesis is not illogical, but non-logical, i.e. outside logic. But once we've formed an opinion, we can expose it to criticism, usually by experimentation; this episode lies within and makes use of logic.
>
> ('Induction and Intuition in Scientific Thought', *American Philosophical Society*, Philadelphia, 1969)

Nevertheless, despite the celebrity endorsements, falsificationism also faces fundamental problems.

Problems

First of all, let's recall what we said above, in the discussion of verificationism: any experimental test of a hypothesis requires 'auxiliary' hypotheses – about the instruments, for example – and this means that the hypothesis we are interested in is not falsified in isolation. When faced with an incorrect prediction it's the whole package that has to be regarded as falsified. But that means that we could always 'save' our hypothesis from falsification by insisting it's the auxiliary hypotheses that are at fault.

Here's an interesting example: neutrinos are elementary particles produced in nuclear reactions, such as those that occur at the heart of, and ultimately power, stars such as our sun (see: https://en.wikipedia.org/wiki/Neutrino). Billions of these neutrinos flow out from the sun every second but they interact so little with matter that we don't normally detect them at all. Furthermore, like the photons that constitute light, they are massless and travel at the speed of light. They do interact very, very slightly with another kind of elementary particle – protons, which, together with neutrons, make up the nuclei of atoms. In order to detect neutrinos, then, you need something rich in protons. Fortunately, we have something like that and it's fairly cheap – washing up liquid!

Now, why would you be interested in detecting neutrinos? Well, as they're produced by the nuclear reactions at the heart of the sun, they effectively give us a way of 'seeing' into this heart. So, if we get a lot of washing up liquid together – and I mean a lot: huge vats of the stuff – then we might see enough interactions with the protons to get an idea of the neutrino flux from the sun. In order to make sure we're seeing only neutrino interactions we'll need to shield our vat from other kinds of particle interactions and one way of doing that is to use the earth itself, by placing the vat down a mine, for example. So, here's a tasty project for some hapless PhD student: sit at the bottom of a mine for months on end, looking at a huge vat of washing up liquid and counting the neutrino interactions (well, using sophisticated instruments to count them).

According to our best theory of how the sun works, we should see a certain neutrino flux rate, that is, a certain number of neutrinos per second. But when the hapless PhD student emerged, blinking into the sunlight, he (or she) reported that the observed rate was one third of that predicted by the theory. Now, if the physicists were true falsificationists, they would abandon their theory, accepting that it had been falsified. However, their theory of the sun had been well confirmed from other observations, drew on and incorporated theories from other areas, such as nuclear physics, themselves well confirmed, and seemed to work for other stars as well as the sun. So they were reluctant to simply drop it, as a straightforward falsificationist would advocate. Instead they began to wonder if their picture of the neutrino was at fault – and it was suggested that perhaps neutrinos come in three different kinds or 'flavours', and oscillate from one to the other as they travel across space between the sun and the tank of washing up liquid, with only one kind interacting appropriately with the protons, so the student only detects one third the rate she should. What the physicists did was to tinker with the 'auxiliary' hypotheses, the extra assumptions that are made when a theory is tested like this. Now if that is just an ad hoc move, then it is hard to see it as more than a face-saving manoeuvre, designed to save the theory at all costs, and if everyone did this when faced with an apparent falsifying observation, then no theory would ever be falsifiable or open to revision because it failed an experimental test and one might wonder just how objective science is. But the change to the auxiliary hypothesis in this case is itself open to empirical testing, to either falsification or confirmation and indeed, the oscillation of neutrinos between the three different kinds has now been confirmed (leading to the award of the 2015 Nobel Prize in Physics; see: http://www.nobelprize. org/nobel_prizes/physics/laureates/2015/press.html), which not only explains the observations of the neutrino flux from the sun but also, it turns out, might explain why there is so much more matter than anti-matter in the universe (see: http://www.bbc.co.uk/news/science-environment-23366318).

Suddenly, then the falsificationist picture and its answer to the question of how science works seem less straightforward. If we can't say for certain that it's the hypothesis we are interested in that is falsified, how can we go ahead and come up with another bold conjecture, how can we learn about the world and how can science progress?

Second, if we look at the history of science, we can find cases where theories face apparently falsifying evidence as soon as they are proposed. They might explain a particular phenomenon but there are others, perhaps comparatively minor or problematic in some way, that conflict with them. In other words, as the philosopher of science Imre Lakatos put it, some (perhaps many) theories are born refuted! But the scientists concerned didn't abandon them and come up with another bold conjecture as Popper says they should. They stuck with their original hypothesis and refused to throw it away. And a good job too, as some of the examples include Newton's theory of gravity and Bohr's famous model of the atom! Consider the former: as soon as it was proposed, it was noted that Newton's Law was in conflict with observations of the moon's orbit, but rather than take his law to be falsified, Newton persisted in developing it and eventually it was determined that the observations were at fault (due the poor accuracy of the instruments).

Finally, it doesn't take much reflection to see that simple falsificationism surely can't be a good strategy to follow. Imagine that not only have you discovered a theory and that it's passed some tests so it's looking plausible, but that it's also the only one in the field. Now imagine that an apparently falsifying observation is made – are you going to give the theory up? Unlikely. The point is, scientists typically don't give up the only theory they have, particularly given the fallibility of observations. Here's Lakatos, again:

> Contrary to naive falsificationism, no experiment, experimental report, observation statement or well-corroborated low-level falsifying hypothesis alone can lead to falsification... There is no falsification before the emergence of a better theory.
>
> (I. Lakatos, 'Falsification and the Methodology of Scientific Research Programmes', in I. Lakatos and A. Musgrave, *Criticism and the Growth of Knowledge*, Cambridge: Cambridge University Press, 1970, pp. 91–196; p. 184)

But there is a more serious problem that afflicts the verificationist and falsificationist approaches alike: How secure are observations? Not very secure at all, according to some, and that's what we'll be looking at in Chapter 6.

Exercise 2

Chapters 4 and 5

Here are some hopefully helpful questions:

Q1 Do you think giving the cause of something is enough to count as an explanation of it? Is it enough in everyday life? Is it enough in science?

Q2 Do you think the advantages of one particular approach outweigh the disadvantages?

Q3 Do you think that explanations should 'get it right'?

Q4 Do you agree that to be meaningful a statement must be verifiable? Can you give any counter-examples?

Q5 Do you think 'survival of the fittest' is a good metaphor for how science works?

Now read this blog post: 'Is Astrology a Science?' (http://www. kuro5hin.org/story/2009/4/10/17459/9222).
   And answer the following questions:

Q6 Why does the author think 'universality' is a problem for verificationism?

Q7 What is the core problem with falsifiability?

Q8 What is the author's view of astrology? Do you agree?

Q9 Read the 'Other Suggestions' section. Can you think of a better criterion for demarcating science from pseudo-science? Or should we simply accept that 'science is merely what scientists do'?

And here are some trickier ones:

A Can you give any examples of explanations that don't fit any of the views outlined in Chapter 4?

B Do you think that different explanatory perspectives can always be integrated? Say why or why not.

C Reflect on the differences between verificationism and falsificationism with respect to their respective employment of inductive vs. deductive logic.

D Why is Kuhn's 'paradigmatic criterion' simultaneously too strong and too weak to demarcate science from pseudo-science?

E In what sense does Quine make it possible for all pseudo-science to be science?

F What is meant by 'epistemological anarchism'?

6

Observation

We recall from the previous chapter both the verificationist and the falsificationist claim that theories and hypotheses are tested against observation statements. So, for theories to be tested (and then either verified or falsified, depending on your point of view) these observation statements must be *secure*. Let me put this another way: a typical view, often expressed by scientists and lay-folk, is that 'science is a structure built upon facts' (J. J. Davies, *On the Scientific Method*, Harlow: Longman, 1968, p. 8), and so we need to ask: How solid are the facts?

Let's begin with a secondary question that will help us get a handle on our first one: How do we get the facts? The answer is obvious: through *observation*; so let's consider the nature of observation, beginning with a common-sense account.

The common-sense view of observation

At the heart of this view lies the claim that the eye is like a camera: light enters through the pupil, is refracted by the lens and an image is formed on the retina. This triggers the rods and cones, and electrical impulses are transmitted by the optical nerve to the brain and, voila, the subject 'observes'. Now, it follows then, that, on this view, two people viewing the same object under the same circumstances – a beautiful sculpture in a well-lit room, say – will 'see' the same thing. Well, actually, no, they won't.

There's more to seeing than meets the eyeball

It turns out that two people viewing the same object under the same circumstances may not, in fact, 'see' the same thing. As the philosopher of science Hanson put it, quite nicely, 'There's more to seeing than meets the eyeball' (N. R. Hanson, *Patterns of Discovery*, Cambridge: Cambridge University Press, 1958, p. 7). Here are some examples:

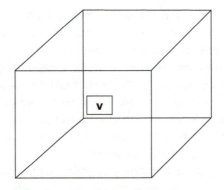

This is the famous Necker cube, named after the nineteenth-century Swiss crystallographer, Louis Albert Necker. If you focus on the vertex labelled 'v', it may seem as if it is at the back of the cube, but a shift in focus can bring it to the front, as if it were projecting out of the page (this is known as 'multistable projection'). The picture is ambiguous and where two lines cross, we are given no clues as to which is behind and which in front. So, two people looking at this image, under the same lighting conditions etc., may 'see' different things. And even then, they only 'see' a cube in the first place because they have accepted and internalized 'Western' art's conventions on perspective (conventions which were only introduced, it is usually claimed, by the Renaissance artist Brunelleschi in the fourteenth century) and the two-dimensional representation of three-dimensional objects. Someone from a different culture operating with different conventions may not see a cube at all, just a bunch of straight lines.

Here's another example, the famous duck/rabbit image (for a vastly better image, see: https://en.wikipedia.org/wiki/File:Duck-Rabbit_illusion.jpg):

Is this a bird of some kind? Or a rabbit? (or some other grotesque creature, badly drawn by the author). Looked at one way it may seem like a bird; looked at another, it looks like a rabbit. The image on the retina is the same, but your perceptual experience and mine may be completely different.

So what you see isn't just determined by the light falling on your retina. It is determined by a host of other factors, by your frame of mind, by your prior beliefs, by my suggestions, for example.

Furthermore, do you remember the first time you looked through a microscope? I do – for a long time all I could 'see' were my own eyelashes! You have to take time and learn how to use a microscope, to distinguish what you're supposed to be looking at from the extraneous material. The same goes for the telescope. Let's recall some history: the telescope wasn't invented by Galileo, but he certainly was one of the first to use it for what we would now regard as scientific purposes, and he developed it further. It was through his crude telescope that he 'observed' the moons of Jupiter and mapped out features on the surface of the moon, observations that (it is typically claimed) did so much to undermine the old, 'Aristotelian' view of the heavens. According to that long-held view, the planets, or 'wandering stars' as they were called, were perfect spheres, held in their orbits by crystalline spheres and driven around their orbits by 'the prime mover', subsequently identified, in the Christian appropriation of

Aristotelian science and astronomy, with God. Galileo's observations disrupted that picture, by apparently showing that Jupiter itself was orbited by subsidiary bodies, its moons, and that the earth's moon is not perfect, but in fact littered with mountains and 'seas' and physical structures.

There's a well-known story about his attempts to convince his colleagues at the University of Padua of the veracity of his observations: Galileo pointed his telescope at Jupiter, invited his colleagues to note the small specks of light that he claimed were moons and more or less stepped back and announced the death of the Aristotelian worldview. But, the usual story continues, his colleagues remained unconvinced, refusing to accept his observations. The noodleheads! How could these supposed learned men refuse to accept the evidence of their own eyes?! The story typically ends with a moral on Galileo's scientific heroism in overcoming the objections of his detractors.

Now, let's probe this nice little story a little further. Were Galileo's colleagues such fools? Here is this man, pointing an unusual instrument at the night sky and asking them to accept that these small specks of light were moons of Jupiter. Could he tell them how the telescope worked? No. Indeed, the theory of optics that would provide such an explanation would not be developed until much later. Could he at least confirm that it magnified objects? Well, Galileo could point it at an earth-bound object, the spire of a church across the way, for example, and show that it magnified that object but, of course, the Aristotelians believed that the laws that applied to earth-bound objects were very different to those that applied to the heavens: that the former suffered death and decay and change in general, whereas the latter is incorruptible and changeless. So why should they accept that something that worked in magnifying objects on the earth would work the same way when pointed at the heavens? And also, even the images of earth-bound objects were not perfectly clear – they were distorted, they suffered from the rainbow effect of chromatic aberration – and so how did Galileo know that these dots of light were not just some optical effect or the result of some defect of the lenses of the telescope?

That colleagues might have been justified in being sceptical of Galileo's observations seems clear if we consider his observations of the surface of the moon. A nice representation of what Galileo drew can be found online: if you compare it with modern photographs, it actually doesn't correspond to what we now see all that well. In particular, Galileo's sketch shows a large crater – which he thought looked a lot like Bohemia – right in the middle of the moon, which simply does not match anything we can currently see (some nice images can be found at http://www.openculture.com/2014/01/galileos-moon-drawings.html). Clearly, Galileo's observations were not that secure.

But let's not be too sceptical. Galileo was able to show that the specks of light he was observing were not defects of his telescope or optical illusions or some peculiar optical phenomena, and he was able to do so quite simply: he observed them on different nights, under different conditions, and was able to show that they shifted position in the sky relative to the fixed position of the telescope. And as he and others used this new instrument more and more they became increasingly adept with it; they began to understand its deficiencies, its flaws, and began to grasp what it could and couldn't do.

This last point is important, perhaps fundamental. Observation involves, in a crucial and profound way, getting a feel for what the apparatus can do, what its limitations are, where it can be 'stretched' and extended into new situations and where it can't. And the observations themselves, or the results of observations, need to be interpreted. Consider a visit to the hospital. The doctor slots the x-rays up into their metal holders in that dramatic way we see on TV and says, 'Ah, here's the problem. Can you see that?' And he points to some blur or blob or shadow that looks little different from the background structures of bones and lungs and organs. Of course you can't see it – you haven't been trained to do so in the way that the doctor has. You need to learn how to see, to observe the fracture, the burst vessel, the tumour. Another scientist turned philosopher of science, Michael Polyani, records a psychiatrist as having once said to his students, 'Gentlemen, you have seen a true epileptic seizure. I cannot tell you how to recognize it; you will learn this by more extensive experience' (M. Polanyi, *Knowing and Being*, Chicago: University of Chicago Press, 1969, p. 123).

Now what's the overall point here? Well, the point is that what you 'see' (the perceptual experiences you have) is not just determined by the image on the retina; it also depends on your training, your experience, knowledge, expectations, beliefs, theoretical presuppositions etc. But then how secure are our observations? Perhaps they are only as secure as the knowledge, beliefs, theoretical presuppositions etc. that inform them. Let's chase this up a little further.

The myth of the secure observation statement

One way in which the apparent security of observations fed into our philosophical view of scientific practice was the following. Philosophers of science focused on the perceptual experiences of observers (which they understood to be private and particular to that individual) and took these to justify observation statements (which are public and can be stated by anyone, obviously) which are, in turn, taken to verify or falsify (depending on whether you're a logical empiricist or Popperian) theoretical statements. Insofar as these are universal – by referring to all bodies being attracted by gravity to one another, for example – they go beyond the observation statements, which typically refer to particular events at particular times and places.

So, now the question is: How secure are observation statements? You might think they're pretty secure, even given what we've considered above. Once we've eliminated any errors and learned how to handle the instruments, surely a statement referring to a particular event is about as secure and objective as anything could be?

Well, consider the following situation. You come home after a long day slaving away on your philosophy course (don't laugh!), ready for a cup of tea and a decent meal, but when you turn on the stove nothing happens and so you call out to your flat mate or partner or casual bystander, 'Hey, the gas won't light' (and no, it's not an electric stove). Is this a secure observation statement? It seems to be – but consider the term 'gas'. This was unknown (at least in the way we use it) until the eighteenth century when Joseph Black separated air into its constituent gases. These are not observable, at least not with the naked eye, and neither is the gas used to cook your pie and chips. 'Gas' is not an observable term, so in what sense is the statement 'The gas won't light' an observation statement?

Here's another example: you and your friend are off to 'Damnation', the Leeds extreme metal music festival, and even though your train doesn't leave until 10 o'clock, your friend insists on getting there half an hour early and spends that time checking the departure announcements. And you sigh and roll your eyes and say to yourself or to no one in particular, 'He's so neurotic.' A simple statement of fact it seems but we know that the concept of neurosis is something that features in certain psychological theories. It's certainly not something observable although the associated behaviour certainly may be. Indeed, we might say that 'neurosis' is a theoretical concept that refers to something unobservable. Likewise when the chemist states that 'The structure of benzene is observed to be ring-like', she is referring to something – the structure of a molecule – that is not observable, but theoretical. And indeed, observation statements are typically couched in the language of theory, or at the very least may involve theoretical terms. How secure are they then? Well, only as secure as the corresponding theory.

Indeed, observation statements that are apparently secure and objective can be false. Consider the following: 'Venus, as viewed from earth, does not change size during the year.' If you observe Venus, also known as the Morning Star at certain times of the year or the Evening Star at others, with the naked eye, it does not appear to change size. But if you observe it with a telescope, it clearly does. The above statement is a *false* observation statement that presupposes the (false) hypothesis that the size of a small light source can be accurately measured by the naked eye, when in fact it can't.

Are there other examples of false observation statements? How about this from Johannes Kepler: 'Mars is square and intensely coloured'! Kepler was one of the greatest astronomers of all time, so how could he make such an apparently elementary error? Well, not to put too fine a point on it, he was in the grip of certain beliefs about how astronomical phenomena should fit into certain geometrical patterns. The problem is that observation statements typically presuppose theory and so they are only as secure as the theory they presuppose.

This obviously raises a problem for the testing of theories and justification in general. Suppose you have a theory of why your stove won't work – you suspect there's something wrong with the mechanism that regulates the gas flow and so to test your theory you fiddle with some valve or other and turn the stove on and then note that 'the gas won't light'. That statement presupposes the concept of gas which ultimately has to be understood in the context of the relevant theory of gases and that theory, particularly as it applies to the behaviour of North Sea gas, is supported by further observations (made far away from your humble flat or bedsit, in the laboratories of British Gas perhaps) which also presuppose theoretical terms, and so it goes on. This is known as the 'experimenter's regress': theory T is tested via some observation O, which presupposes some further theory T', supported by observations O', which presupposed theory T'' and so the regress goes on. The problem is, where is the bedrock? On what does the security of observations rest?

One response is just to swallow this apparently unpalatable point. Observations do not rest on any bedrock and observation statements are fallible too, just like the theories they support or falsify. Popper, again, put it very nicely:

> The empirical basis of objective science has nothing 'absolute' about it. Science does not rest upon a bedrock. The bold structure of its theories rises, as it were above a swamp. It is like a building erected on piles. The piles are driven down from above into the swamp, but not down to any natural or 'given' base; and if we stop driving the piles deeper, it is not because we have reached firm ground. We simply stop when we are satisfied that the piles are firm enough to carry the structure, at least for the time being.
> (K. R. Popper, *The Logic of Scientific Discovery*, London: Hutchinson, 1968)

But, you might insist, setting all of these concerns to one side, unbiased observers can at least agree on which observation statements to accept. Or can they?

The myth of the unbiased observer

There is a well-known and widespread view that scientific observation should be unbiased and free from presuppositions. It is this lack of bias that helps to underpin the objectivity of science. But let us ask the question: *Should* observers be unbiased? Just think about a botanist, say, who heads out into the field – the Amazon jungle, say – to conduct some botanical 'observations'. Is she just going to parachute into the jungle, unbiased and without presuppositions and simply start observing, left, right and centre? And observing what? All the plants, all the trees, all the strange animals and insects? No, of course not. She will know what she is looking for, what counts and what doesn't, what conditions are relevant etc.; she may even have, and will typically have, some theory in mind, some set of hypotheses that is being tested. Of course, serendipitous observations happen, new plants or animals are discovered, for example, but a botanist observing without bias in the field would be overwhelmed.

Here's another, more historical but actual, example. One of the great theoretical advances of the nineteenth century, indeed of all time, is Maxwell's theory of electromagnetism, which effectively unified electricity and magnetism, explained light as an electromagnetic phenomenon, and which predicted the existence of radio waves. More specifically, it predicted that these waves should travel at the same speed as that of light. In 1888, the great German scientist Hertz attempted to detect radio waves and observe whether the speed was indeed that of light. Now, if he had made his observations in an unbiased manner, free from presuppositions, what would he have observed? Would he have observed the readings on the meters, the colour of the meters, the size of the meters, the size of the laboratory, the size of his underpants?! Of course not: many of these 'observations' would be irrelevant, but what determines which observations are relevant and which are not? The obvious answer is that theoretical presuppositions help to sort this out. Indeed, when Hertz made his observations, he found that the speed of radio waves did not equal the speed of light because something he did not regard as relevant turned out to be very relevant indeed. Despite all his efforts, his observations refused to confirm Maxwell's prediction.

Now, if Hertz had followed Popper's falsificationist approach he would have declared Maxwell's theory falsified and thrown it out. But this theory was extremely successful in explaining other phenomena and Hertz believed that it was his observations that were at fault. In fact, it was discovered after his death that one of the factors listed above, which one might think would be irrelevant, turned out to be very relevant indeed. No, it wasn't the size of his underpants but the size of his laboratory. It turned out that the radio waves generated by Hertz's apparatus were bouncing off the laboratory walls with sufficient intensity to interfere with the waves being observed and that's why he kept getting the wrong value for their speed.

Here we see how observations are guided by theory. That in itself seems unproblematic, reasonable even. But what if different observers have different theoretical presuppositions? The conclusion would seem to be that they will 'observe' different things. Here's an example from psychology: consider someone who falls in love, but considers him/herself unworthy as compared with the virtues of the object of their adoration. The Freudian observes a narcissist, who has displaced that portion of their libido, which attaches to the 'Super-Ego', by the standards of which the 'Ego' is judged unworthy. The Adlerian, on the other hand, observes someone compensating for their own perceived inferiorities by projecting onto the object of adoration those qualities in which they feel they are deficient. Different theoretical backgrounds, different observations.

The overall conclusion, then, is that theory plays a number of roles in observations – it guides them, informs them, gives meaning to the statements reporting observations – and the generic philosophical claim covering this situation is that observation is 'theory-laden'. More important than what it's called, however, is what it implies for the objectivity of science. It might seem that observations are not very secure at all and if science is a structure built on facts it's a bit rickety. In the next chapter, we will see that we should not be too pessimistic and that there are ways in which the security of observations can be underpinned.

7

Experiment

Introduction

Let's recall what we've covered in the last chapter. First, what you 'see' (that is, the perceptual experiences you have) is not just determined by the image on the retina: it also depends on your experience, knowledge, expectations, beliefs, theoretical presuppositions etc., which help you select what's relevant, what's real, what's an artefact and so on. Second, the role of *instruments* in observation is crucial. That's perhaps an obvious point but is one that's often overlooked in certain philosophical discussions. And finally, observations are often guided by theory (we recall the example of Hertz and his frustrated search for radio waves). We're already beginning to see that observation in science, and hence the process by which theories are justified, is a bit more complex than we initially thought.

In particular, what these points show is that observations and experimental results in general are *revisable* (we recall Popper's piles!). But if that is the case, what about the secure observational base for science? Again, we recall that this is not just a philosophical assumption, built into accounts of justification, but underpins the 'common-sense' view that science is built on 'facts'. Well, the short, sharp response to this concern is that science doesn't need such a base! But then what about objectivity? How can science be objective if it's built on shifting sands instead of the bedrock of secure observations? That's going to be the main focus of this chapter and we'll also be looking at how the justification of theories is further complicated through the introduction of models.

The objectivity of observation

So, here's our next question: If observations are revisable, at least in principle, how do we ensure that they are as secure as they can be? And in particular, how do we distinguish 'genuine' observations from those that are mere artefacts, the result of a flaw in the instruments, perhaps, or poor observational technique?

Let's recall Galileo and his famous telescope and ask ourselves, how did Galileo convince everyone that the 'starlets' around Jupiter were real and not artefacts, when he had no theory of the telescope and couldn't explain to his critics how it worked? Well, the answer is pretty straightforward: he pointed the telescope somewhere else and the starlets vanished (no kidding!). He observed the starlets over a period of time and noticed a consistent and regular pattern of movement. That is, daily measurements of the movements of the starlets generated a consistent and repeatable set of data, leading Galileo to conclude that these specks of light were not an illusion. Furthermore, his observations were repeated elsewhere, and (later) using different *kinds* of telescopes, and his conclusions confirmed.

Here's another example: suppose we are interested in observing the structure of red blood cells. Using the most sophisticated instruments we have available, such as an electron microscope, we 'see' a configuration of dense bodies (and of course electron microscopes work on very different principles from optical microscopes, so what we 'see' is actually an image generated by a very complex process). Now, in order to make such an observation, we don't just stick some blood on a slide and place it under the microscope. Biological samples such as this typically have to be prepared in certain ways, in order to stabilize them, in order to highlight the features we are interested in and so forth. A tissue sample might be 'stained' with certain chemicals for example and the question naturally arises, Is what we observe an artefact of the instrument or the way the sample is prepared?

One way to answer this question is to mount the sample on a grid and observe with a different kind of instrument. So, we might take our blood sample and examine it with a fluorescent microscope, which operates on very different physical principles from the electron microscope. If we see the same configuration of dense bodies then we are entitled to conclude that this is not an artefact but a genuine feature of the red blood cells. Taking our tissue sample, we might use different techniques to stain it or fix it. Again, if we see the same structures, using all these different techniques, then we might feel inclined to accept the structures as really there, and not some artefact of our preparations. In fact, scientists use a variety of strategies to validate their observations: an obvious one has to do with checking and calibrating the equipment by observing if it reproduces or accurately portrays known phenomena, before deploying it to observe the unknown. Of course, there may be subtle issues associated even with this obvious practice. Think back to Galileo and his attempt to convince his colleagues of the veracity of his observations by pointing the telescope at familiar earth-bound objects and showing that they were simply magnified – if you believe that earth-bound and heaven-bound objects obey different laws, you're hardly likely to be convinced!

But there are other strategies at hand. You might use your equipment to reproduce artefacts already known to be present and which can then be accounted for. For example, if you are observing a sample suspended in a solution, by looking at the spectrum of light given off when the sample is irradiated, say, then you would expect to see also the known spectrum given off by the solution. Seeing that would indicate that your equipment is working correctly. Another fairly obvious move is to eliminate or, at the very least, allow for as many sources of error as possible. Anyone who has done any experimental work in science will appreciate that one of the most boring but also most essential aspects of the job is to identify, eliminate or account for (using statistical techniques) the many sources of experimental error that can feature in even the simplest experiment. Often more time will be spent doing this than actually running the experiment!

Another technique is to use the regularity of the results to indicate that the observations are secure. If you see the same thing, day in day out, under different circumstances then chances are that you are not seeing an artefact or random 'blip' but something actually there. Even more importantly, if you see the phenomenon in question behaving in a regular fashion then that can be one of the most important indicators that you've actually got something and that your observations are secure. Again, recall Galileo and the moons of Jupiter – it's hard to argue that the little dots of light he was seeing are just defects of the telescope lenses when they move around the planet in a regular and, ultimately, predictable fashion! Of course, this becomes problematic if the phenomena we are concerned with are transient or difficult to reproduce, which is why claims regarding such phenomena are often scientifically controversial. Finally, we can use theory itself: if our apparatus is based on well-confirmed theory then that provides further reason for supposing that it is working well and that we are observing a legitimate phenomenon and not some artefact or blemish. Indeed, we might use this theory to identify and account for such artefacts and blemishes. As I mentioned, at the time Galileo made his observations, the theory behind the telescope was not known but eventually its operation was explained in terms of the wave theory of light and that could then be used to further explain and account for the various defects and distortions that the apparatus is prey to. Likewise, radio telescopes have for many years been used to probe the heart of the galaxy and beyond but the theory of their operation is now very well known and although anomalies do occur the theory itself, in addition to these other strategies mentioned here, can be used to identify and distinguish what is an artefact or error and what is a genuine observation (an excellent overview of these strategies can be found at http://plato.stanford.edu/entries/physics-experiment/).

So, what's the conclusion? First, philosophers have tended to think of observation as more or less simply a matter of opening your eyes! You might need to get the lighting right and generally adjust the ambient conditions but basically, on this view, you just open your eyes and observe what is in front of you. But if we consider what actually goes on in science, and particularly modern science, we see (ha, ha) that observation is not a mere passive operation but involves an *active* engagement with the world. Most, if not all, observations in science are made via *experiments* and experiments involve not just passively *representing* the world but actively *engaging* with it and in some cases *intervening* in it, by, for example, preparing the sample, staining to highlight the features you are interested in and so on. We ensure an observation is genuine by eliminating artefacts and illusions, obtaining repeatable data etc. and it is on this basis that the objectivity of science is secured. So, the lesson here is that objectivity is not something that is guaranteed by merely opening your eyes and observing what is in front of you, it is a *practical achievement* (a very nice discussion of this can be found in A. Chalmers, *Science and Its Fabrication*, Maidenhead: Open University Press, 1990).

But what about the point we noted in the previous chapter, that observations are frequently guided by theory, that statements about observations may contain terms that are theoretical and in general that it is difficult, perhaps impossible, to nicely separate theory from observation in the way that some people have imagined. The question now is does this 'theory-ladenness' of observation undermine the testing and justification of theories? The obvious answer is, no, not if the theory which is loading the observation is different from the theory being tested. If we use our telescope to test some theory about astronomical phenomena, then the observations we make will be 'laden' with the theory of optics we have to assume is correct in order to help eliminate artefacts and aberrations and generally validate the observations. Indeed, as I said earlier, the theory of the apparatus can help us to secure the observations we make. If we were using the telescope to make observations bearing on the relevant part of the theory of optics itself (the part that is used to explain the working of the telescope), then it would be a different matter of course. Then we would just be running round an experimental circle and there is precious little security to be found in that. But what if some kind of observational chain is set up in which we use an instrument to test a theory and the relevant observations are effectively laden with some theory that was tested via observations which were laden with the first theory we were trying to test? That would be a longer but still nasty loop which would clearly undermine the objectivity of the test; however, scientists are generally pretty adept at avoiding such vicious circles and there are few if any that occur in practice. Of course, the bigger the loop, embracing theories apparently quite 'distant' in some sense, the harder it may be to detect.

Top down vs. bottom up

Let's go back to our picture of discovery and justification in science: we begin with a hypothesis, discovered perhaps via some set of heuristic moves, and we deduce some prediction from it (as discussed in Chapter 4). We then set up our instruments, prepare our sample and make our observations. If the data confirms our prediction then it's Nobel Prizes all round once again; if not, then we either throw our hypothesis away and go back to the drawing board, if we're hard core falsificationists, or, more likely, check our instruments, check our experimental technique, maybe tweak the hypothesis a bit and so on.

Indeed, in some cases the test of our hypothesis might be so important that it comes to be described as an 'experimentum crucis', which is just Latin for 'crucial experiment'. These experiments come to be called that because they are seen as decisively coming down on the side of one theory or hypothesis rather than another. So, an example of this would be Eddington's observation of the bending of starlight around the sun (we'll leave to one side the point that this wasn't really an *experiment* as such), which, as we noted, was taken to be crucial in supporting Einstein over Newton when it came to which was the correct theory of gravity.

Of course, as we have also noted, the impact of experiments and observations can be moderated, or even diverted, by the role of auxiliary hypotheses and assumptions used to obtain the relevant predictions. In Eddington's case, there were comparatively few such assumptions that could be easily questioned and that helps explain why his confirmation of Einstein's theory was accepted so readily. But we could imagine someone seeking to defend Newton's theory by adding an extra hypothesis about the effect of gravity on light within the Newtonian framework in an attempt to account for the observations. That would be deemed to be 'ad hoc' in the sense of being deployed solely for the purpose of saving the theory and such a move is typically frowned upon by both scientists and philosophers. And of course this extra hypothesis would have to be independently tested itself to count as helping to save Newton's theory (it turns out that if light is regarded as a particle, as it is in a sense in quantum physics, then Newton's theory does predict the bending of starlight, but only by about half the amount observed).

Whatever we do with the hypothesis and however we regard the observation or experiment, the underlying structure of this picture is as follows:

And it might be thought that this sort of relationship is reflected within the scientific community itself: that is, the theoretical folk come up with their hypotheses, deduce their predictions and they basically send the message down to the lab: 'Yo ho, lab minions – test this and tell us what you observe'! Does this picture of theory testing accurately represent what goes on? The short answer is again no. This is very much a top down picture according to which science is basically theory driven, and actual practice is more complex than that (and actual experimental scientists might well point out that they are not the theoreticians' trained monkeys!). Indeed, it has been suggested that experiment has a life of its own, one that is independent from theory, and that this life has been a bit of a secret as far as philosophers have been concerned.

The secret life of experiment

Consider the following statement by Justus von Liebig, one of the great nineteenth-century chemists (famous for his discovery of nitrogen as a plant nutrient and – this is one for the Brits among you – for founding the company that made Oxo cubes):

> Experiment is only an aid to thought, like a calculation: the thought must always and necessarily precede it if it is to have any meaning. An empirical mode of research, in the usual sense of the term, does not exist. An experiment not preceded by theory, i.e. by an idea, bears the same relation to scientific research as a child's rattle does to music.
>   (J. von Liebig, 1863, quoted in I. Hacking, *Representing and Intervening*, Cambridge: Cambridge University Press, 1983, p. 153)

Now, that's quite a strong line to take (especially the comparison of experiment with a child's rattle!) and it raises the question: Do theories always come first? The answer is no, not necessarily – experiment may have a life of its own. This 'secret' life has been very nicely explored and captured by Ian Hacking in his book, *Representing and Intervening*.

He notes, first of all, that observations and experiments may precede theory. So, he gives as examples, Rasmus Bartholin's discovery of double refraction in Iceland Spar (a form of calcite – a nice image can be found at http://geology.about.com/library/bl/images/blcalcite.htm); Christiaan Huygens' observation of polarization in Iceland Spar; the diffraction of light observed independently by Francesco Grimaldi and Robert Hooke; William Herschel's observation of fluorescence; the Rev. Brown's observations of Brownian motion (the minute jiggling of pollen grains suspended in a liquid, explained (by Einstein) through atomic collisions); Henri Becquerel's discovery of the photoelectric effect (in which electromagnetic radiation of the right frequency causes electrons to be emitted by metals – the basis for the solar cell).

In all these cases, the theory accounting for these observations came later. Just as zoologists occasionally still stumble across previously unknown species of mammals in the jungle, so scientists in general may discover some new effect or phenomenon. Of course, let's not forget the lessons of Chapter 3: these discoveries are typically made by well-trained, observant scientists who are able to recognize that they've come across something significant.

So Bartholin, for example, was the son of a doctor and a mathematician and not surprisingly, perhaps, went on to become both a professor of geometry and a professor of medicine. He not only explained the peculiar phenomena associated with Iceland Spar in terms of refraction but also made numerous astronomical observations and advocated the use of quinine in tackling malaria. Likewise, Becquerel came from a distinguished family of scientists (his father did work on solar radiation and on phosphorescence) and is perhaps best known for his discovery of radioactivity: following a discussion with Henri Poincaré on the newly discovered x-rays, he decided to see if they were connected to phosphorescence, using some uranium salts that were known to phosphoresce. The exposure of a covered photographic plate when placed near the salts led to the realization that another form of radiation was at work here. In 1839, aged only nineteen, he was studying the chemical effects of light when he noted that when light of a certain frequency was shone on electrodes immersed in a conducting solution, a current was produced. These were not guys who simply stumbled over these observations! Granted, they were made before the relevant theories were in place to explain them, and in that sense they had their own experimental 'lives', still, they were not made outside of any scientific and, in particular, theoretical context.

Sometimes, of course, the effect may languish for years without a theoretical explanation but sometimes not. Both Brownian motion, for example, and the photoelectric effect were not explained until the beginning of the twentieth century – by the same person, as it happens, Einstein again (as I noted in the last chapter, he got the Nobel Prize for his account of Becquerel's discovery). However, there are also cases where a phenomenon is discovered and then related to a particular theory almost immediately. Hacking gives a very nice example of this where the theory and observation meet up unexpectedly:

> Some profound experimental work is generated entirely by theory. Some great theories spring from pre-theoretical experiment. Some theories languish for lack of mesh with the real world, while some experimental phenomena sit idle for lack of theory. There are also happy families, in which theory and experiment coming from different directions meet.
>
> (I. Hacking, *Representing and Intervening*)

In the 1950s, there were two competing cosmological theories about the origins of the universe. One posited that it began with a 'big bang', from which not just matter, but space-time itself, emerged and expanded. That expansion continues, as the galaxies move away from each other (imagine taking a balloon, drawing dots of ink on it and then blowing the balloon up). The alternative theory states that as the universe expands, more matter is created to take its place, so the universe as a whole remains in a steady state. At the time, there was no further evidence that could discriminate between these theories.

Completely independently, two scientists were experimenting with microwaves at Bell laboratories in the US. While trying to make their observations, they noticed a background hiss, in the microwave range of frequencies, and no matter what they did, they couldn't seem to track down the source. They eliminated a range of possibilities, including local radio and TV stations. They even wondered if it might have something to do with the heat generated by the pigeon poop in their large, outdoor microwave emitters and receivers, so they went out and shovelled it away (there's dedication to science for you!). And then around this time, a cosmologist at Princeton, Robert Dicke, suggested that if there had been a 'big bang' it should have left a residue of low-level microwave radiation throughout the universe. He had been looking for evidence of this when Arno Penzias and Robert Woodrow Wilson got in touch, leading to his remark to his colleagues that, 'we've been scooped' (a nice account of this 'meshing' of observation and theory, including a recording of what the remains of the big bang sounds like, can be found at http://www.npr.org/templates/story/story.php?storyId=4655517).

Subsequent observations have shown that the background radiation is not completely uniform, but contains irregularities which explain why matter began to clump together, forming galaxies, stars, planets and us. A touch melodramatically, George Smoot, one of the pair of physicists who made these latest observations, and received the Nobel Prize for them in 2006, remarked that it was like 'seeing the face of God' (an even more detailed map of this 'face' is being put together by the European Space Agency's 'Planck Mission': http://www.bbc.co.uk/science/space/universe/exploration/planck_mission/).

So, we can begin to see that the relationship between theory and experiment is not always theory driven, not always 'top down'. But we can explore this relationship even further. Theories are often highly complex, whereas observations are typically much simpler. In order to get from one to the other we need to pass through an intermediate stage and this generally involves the construction of some kind of model. We've already looked at the role of models in discovery, but now we shall see that they are important for justification as well.

The role of models in science

Let's begin with the question: How is theory brought to mesh with experiment? This is an interesting question because a theory will (surprise, surprise) contain theoretical terms referring to unobservable entities, such as electrons, genes and so forth, whereas the results of experiments are expressed in terms of observation statements, containing observation terms which refer to observable objects, like meter readings, chemical precipitations, or whatever. How are these two different kinds of statements to be related? One answer is that we need a kind of *dictionary* to relate the theoretical terms of the theory/hypothesis with the observational terms of obser- vational statements, so we can then deduce the experimental or observational consequences. This dictionary will contain 'bridge' statements that tie the theoretical statements to observational ones by summarizing experimental procedures: that is, they say, 'if you do such and such, set up the following experimental arrangement, you will observe the following result'. Now, what this means is that a theory, strictly speaking, must consist of more than just theoretical statements – in order for it to be related to observation, it has to include these bridging statements as well. So, such statements are constitutive of the theory; they are part of its fabric.

However, this has an unfortunate but quite obvious consequence: if the bridge statements are a constitutive part of the theory, and these statements summarize experimental procedures required to tie the theoretical statements of the theory to observational results, then if we introduce a new experimental procedure, as often happens in science, that has to be represented in another bridge statement, which means that, strictly speaking, we have a new theory, because the constitutive parts have changed. The obvious way out of this problem is to say that how the theoretical statements relate to observational ones is expressed 'outside' of the theory, somehow. Also, this whole approach presupposes that observational state- ments contain only observation terms, but as we've seen this may not be so. And, this doesn't seem to be how scientists themselves operate – they just don't relate theoretical and observational state- ments in this way. How do they relate them? Via models.

That is, scientists *don't* simply deduce experimental/observational consequences; they construct models that 'mediate' between theories and the observations. There are a number of reasons why scientists will proceed in this way but one is that theories are often quite complex and difficult to work with. So a scientist may build a simplified model, containing significant idealizations that allow the scientist to ignore certain factors, for example, and easily relate the theory to observations.

What is a model, then? It may be a physical construct, built out of wire and tinplate, for example, as in the case of Crick and Watson's model of the helical structure of DNA that we considered in Chapter 3. Models may also be conceptual and mathematical. We recall the 'billiard ball' model of a gas, in which the atoms of the gas are represented by billiard (or snooker) balls. In other words, the billiard balls are taken to be an analogy for the gas atoms. But if this were all there is to the model, it would be pretty uninteresting. Since the billiard balls are subject to Newtonian mechanics – that is, their motion across the billiard table is described by Newton's laws – the model encourages us to apply these laws to the gas atoms too. In this way, something comparatively unfamiliar – the motion of gas atoms – comes to be understood in terms of something more familiar (at least for those of us who have wasted our youth in the snooker hall).

We also recall the liquid drop model of the nucleus, which does more than offer a useful picture of the nucleus; just as in the case of the billiard ball model, it encourages the transfer of the equations that apply to the liquid drop to the nucleus. Of course, there are limits to the applicability of this model – ultimately the nucleus is a quantum beast and has to be explained by quantum physics. Nevertheless, such models are useful heuristically – they may help scientists to develop more complex and sophisticated theories – and they allow scientists to obtain experimental results easily. So, by constructing a liquid drop model of the nucleus, scientists can obtain predictions of how the nucleus will behave under certain conditions, even when the theory is horribly complicated.

This sort of example can be generalized and models can be found in many domains of science: computer models of the brain; 'circuit' models of ecosystems; or 'neural network' models of 'protein domain evolution'. Typically, theories are just too complex for us to straightforwardly deduce experimental consequences, so we construct a simpler, mathematically tractable model that, on the one hand, contains idealizations of certain aspects of the theory and, on the other, captures at least some aspects of the relevant phenomena. In this way, the models can be said to 'mediate' between theory and phenomena and may become so important that they themselves become the focus of scientific activities, rather than theories themselves. Here's how one commentator puts it:

> The core of my argument involves two claims (1) that it is models rather than abstract theory that represent and explain the behaviour of physical systems and (2) that they do so in a way that makes them autonomous agents in the production of scientific knowledge.
>
> (M. Morrison, 'Models as Autonomous Agents', in M. Morrison and M. Morgan (eds), *Models as Mediators*, Cambridge: Cambridge University Press, pp. 38–65; p. 39)

The first part of this claim is apparently straightforward. Of course, models represent and explain, at least to some extent. Consider the liquid drop model of the nucleus yet again. This represents the nucleus, obviously, as a liquid drop and, at least, helps to explain the behaviour of nuclei, such as fission, through the analogy of a drop splitting apart. We've discussed explanation in science in Chapter 4 and will touch on it again in Chapter 8 but we'll just note here that explanation via models offers a further dimension to this notion. Representation in science is also a topic that has attracted a lot of attention recently. Just as philosophers of art try to come up with accounts of how paintings, for example, represent, so philosophers of science have begun to consider how theories and models represent. You might think that's a pretty easy question to answer: a painting represents its subject through being similar to the subject. Leaving aside issues to do with abstract art, we can see how that works: Vincent van Gogh's famous *Sunflowers* represents through features of the painting – the yellow of the petals, the shape of the flowers etc. – being similar to features of real flowers. And it might seem that we can export this kind of account across to science: theories and model represent via similarity. So, the model of the simple pendulum represents an actual pendulum since the bob in the model is similar to a real bob and so on. Likewise, the liquid drop model is similar to a real nucleus, in certain ways.

However, the problem with that sort of account is that similarity cuts both ways: certain features of the flowers are similar to the relevant features of the painting, but we would not say that the flowers represent the painting. Likewise, we would not say that a pendulum represents the model or the nucleus represents the liquid drop. This sort of argument has led some philosophers – of both art and science – to come up with alternative accounts of representation, or, at the very least, to modify similarity type approaches. Some have even suggested that perhaps we need to be wary of importing considerations of how art works into our description of how science works!

The second part of the above claim about models might appear weird. In what sense can a model be an 'agent', autonomous or not? Agents have intentions, but models certainly do not. What is meant here, however, is fairly straightforward: models may become so important in a particular science, or field of science, that they become the focus of attention and the locus of activity, in the sense that rather than high-level theories doing the explaining and representing, it is the lower-level models. They can be regarded as 'functionally' autonomous, in the sense that in terms of these representative and explanatory functions they are comparatively independent of theory within scientific practice and further models may be developed from them, so that scientific progress itself proceeds at this level. Some have argued that this means that theory can be left out of the story but that is to go too far; typically, the kinds of interesting models scientists are concerned with draw on a range of theoretical resources in their development. However, as interesting as this discussion is, it takes us a little too far from the theme of this chapter, so let's get back to that.

What we have is the following sort of picture:

Theories
↓
Models
↓
Observation

How does this affect our view of the justification of theories? Well, we can see how our answer will go: the 'mediation' of models between theories and observation means they act as a kind of buffer when it comes to justification. The falsifying force of an observation is going to be blunted by the model since it may be the idealized elements that are at fault, rather than anything to do with the theory. This obviously makes justification more complicated, as the impact of observations becomes unclear. But now let's look in the other direction: theories are supposed to explain phenomena, but what are phenomena? And how are they related to observations and data?

What do we mean by a 'phenomenon'?

Here's the Oxford English Dictionary's definition of 'phenomenon':

> a thing that appears or is perceived, esp. thing the cause of which is in question; (Philos.) that of which a sense or the mind directly takes note, immediate object of perception; remarkable person, thing, occurrence, etc. [f. LL. f. Gk *phainomenon* neut. pres. part. (as n.) of *phainomai* appear (*phainō* show)]

The traditional view of scientific phenomena is that they are observable, remarkable, require explanation and are discovered in nature. Examples include rainbows, lightning, the bending of starlight around the sun, crowd behaviour ... Phenomena are all over the place! But here's a different view.

Constructing phenomena

According to Hacking (see again: I. Hacking, *Representing and Intervening*, Chapter 13), phenomena are typically *created* not discovered. Now, that's quite a radical idea. As far as scientists are concerned, phenomena are public, regular, obey certain laws and are often associated with unusual, perhaps exceptional, events. Truly exceptional, instructive phenomena are sometimes called 'effects', such as the photoelectric effect, recency and 'chunking' effects in memory, and so on.

Hacking claims that such effects do not exist outside of certain kinds of apparatus. As embodied in such apparatus, the phenomena 'become technology', to be routinely and reliably reproduced. So, if the right arrangement exists, or is put together, the effect occurs, but nowhere outside the lab do we generally find the 'right' arrangement. Here's Hacking again:

> In nature there is just complexity, which we are remarkably able to analyse. We do so by distinguishing, in the mind, numerous different laws. We also do so, by presenting, in the laboratory, pure, isolated, phenomena.
>
> (ibid.)

So, on this view, in nature there are few phenomena just sat there waiting to be observed; most of the phenomena of modern science are manufactured, or constructed. In other words, phenomena are created by *experiment*:

> To experiment is to create, produce, refine and stabilize phenomena. If phenomena were plentiful in nature, summer blackberries there just for the picking, it would be remarkable if experiments didn't work. But phenomena are hard to produce in any stable way. That is why I spoke of creating and not merely discovering phenomena. That is a long hard task.
>
> (ibid., p. 230)

From this perspective, the usual emphasis on the *repeatability* of experiments is misleading. Instead, experiments are *improved* until regular phenomena can be elicited. This seems a very restricted view of phenomena but Hacking seems to have a point. The kinds of observable phenomena mentioned above are certainly not repeatable or easily controllable. They may have some significance in the early days of science – just as Franklin's work with lightning was important in the early days of the study of electricity – but serious scientific work proceeds once we can isolate the phenomenon, repeat it, control it and investigate it. And typically, the phenomena investigated at this stage are not observable, or at least not straightforwardly.

Here's another, more sophisticated view of phenomena that reflects this last idea.

Finding phenomena

The central idea of this more sophisticated view is that phenomena and data should be carefully distinguished insofar as the former are not observable, unlike the latter, and the data support phenomena rather than theories:

> ...well-developed scientific theories ... predict and explain facts about phenomena. Phenomena are detected through the use of data, but in most cases are not observable in any interesting sense of that term.
>
> (J. Bogen and J. Woodward, 'Saving the Phenomena', *The Philosophical Review* 12 (1988): 303–52; 306)

So, data serve as evidence for the existence of phenomena, and the latter serve as evidence for, or in support of, theories. Theories in turn explain phenomena and not data. So now our picture looks like this:

Theories

↓

Phenomena

?

Data

But here's the obvious question: how are 'data' distinguished from 'phenomena'?

> Examples of data include bubble chamber photographs, patterns of discharge in electronic particle detectors and records of reaction times and error rates in various psychological experiments. Examples of phenomena, for which the above data might

provide evidence, include weak neutral currents, the decay of the proton, and chunking and recency effects in human memory.

(ibid.)

Data are straightforwardly observable and specific to particular experimental contexts. They are the result of highly complex arrangements of circumstance and they are not only relatively easy to identify and classify, but are reliable and *reproducible*. Phenomena, on the other hand, are not observable, not specific, have certain, stable and repeatable characteristics and in general are constant across different experimental contexts.

How do we get from data to phenomena? Phenomena are inferred from data and the robustness and objectivity of phenomena are underpinned by the reliability of data. And the reliability of data is typically established by the experimental method, by empirically ruling out or controlling sources of error and confounding factors, or by statistical arguments and so on. How are phenomena related to models?

When we talk about 'inferring' the phenomena from the data, what we're actually talking about is inferring *properties* of the relevant objects, processes or whatever. These properties feature in an appropriate model of the objects, processes, etc. concerned and this model will be a *model of the phenomena*. Here we have a different kind of model; not a model obtained from a theory by simplifying things or via idealizations because the theory is too complex: this is a model built from the bottom up. Likewise, the data are represented by 'data models' and the models of phenomena are supported by the relevant data models, so we obtain a hierarchy of models:

[t]he concrete experience that scientists label an experiment cannot be connected to a theory in any complete sense. That experience must be put through a conceptual grinder that in many cases is excessively coarse. Once the experience is passed through the grinder, often in the form of the quite fragmentary records of the complete experiment, the experimental data emerge in canonical form and constitute a model of the experiment.

(P. Suppes, 'What is a Scientific Theory?', in S. Morgenbesser
(ed.), Philosophy of Science Today, New York: Basic Books, 1967,
pp. 55–67; p. 62)

The 'raw' data are fed through this conceptual grinder to generate a model of the experiment, or data model, and from such models the 'phenomena' are inferred. This offers a much more sophisticated view of the relationships between theory, data and phenomena, one that can be investigated further; but we'll leave it there for now.

Conclusion

What can we say about justification? Well, it's more complicated than we thought! Simple prescriptions like 'Verify!' or 'No, falsify!' don't capture the complexity of scientific practice where, sometimes, one 'stunning' verification is enough, but one apparently dramatic falsification is not. The security and objectivity of the data is more problematic than we thought; observation is apparently 'theory-laden'. Nevertheless, once we acknowledge the active nature of observation, we can acquire a measure of objectivity. The relationship between theory and data isn't just 'top down'; there is a sense in which experiment may 'have a life of its own'. The relationship between theory and data may be 'mediated' by models and the phenomena explained by theories are inferred from the data and represented by models in complicated and sophisticated ways. Accommodating all of this is a difficult job, but, if we're going to describe justification in science accurately, it's one we have to tackle as philosophers of science.

Exercise 3

Based on Chapters 6 and 7 here are some questions to prompt further thoughts:

Q1 Think of some examples where direct observation may be unreliable. Are these examples relevant to analysing observation in science?

Q2 Consider an example in which our beliefs can influence what we see. Is this example relevant to what happens in science?

Q3 What does 'robustness' mean in connection with (scientific) observations? How is it achieved?

Q4 What does it mean to say that observation is 'theory-laden'? Give some examples.

Q5 How do *models* 'mediate' between theories and observations? Give an example.

Q6 What is the distinction and relation between *data* and *phenomena?* Come up with some examples to help explain the distinction.

And here are the more advanced ones:

A How is robustness related to scientific induction?

B Does 'the theory-ladenness of observation' undermine the objectivity of testing scientific hypotheses?

C Would a Popperian have a problem with the 'model approach' to analysing how scientific theories are tested in experiments? Why?

D   What do Bogen and Woodward mean by saying that phenomena are 'stable'?

The following might be useful: J. Bogen, 'Theory and Observation in Science', *Stanford Encyclopaedia of Philosophy*, http://plato.stanford.edu/entries/science-theory-observation/

8

Realism

Introduction

So, you've discovered your hypothesis and you've used it to explain some things, and also subjected it to rigorous testing, taking into account all that we've said so far, and it seems to be holding up in the face of all the evidence. Does that mean that what it says about the world is true? Does that mean that the objects and processes it presents to us actually exist? The obvious answer would be to say, 'Yes, of course' and if you are inclined to adopt that line, then you are a 'realist' of some stripe or other. Now it might seem the most obvious answer but as we'll see, objections can be raised to it. Those who raise such objections are known as 'anti-realists' and again, as we'll see shortly, they too come in different forms.

So, this is the fundamental question for this chapter: What do scientific theories tell us? Here are three different answers:

A1: They tell us how the world is, in both its observable and unobservable aspects (realism)

This is the realist answer. Realists take theories to be true, more or less, and tell us how the world is, not just with respect to what we can observe, but also when it comes to unobservable features as well. Now, drawing the distinction between the observable and the unobservable is a bit tricky. First, do we mean 'observable' with the naked eye or with scientific instruments? Scientists themselves adopt the latter understanding and talk of observing biological processes, molecules, even atoms. But then even if you're happy with talk of observing microscopic bugs through an optical micro-scope, I'm willing to bet you're less comfortable about observing clusters of atoms through a scanning electron microscope. In the former case we have a set of intervening lenses between our eyes and the sample; in the latter, we have a much more complex set-up of electrical devices, not to mention the computer enhancement involved.

Now you might say that it shouldn't matter how the observing instrument is constructed and that we simply can't draw a sharp line between those devices which contribute to 'genuine' observations and those which do not. If you say that, you might be inclined to go one of two ways: either it doesn't matter whether you use instru-ments for observation; or it does, and naked-eye observation is the only form that counts. Even if you adopt a hard line and go with the latter option, things are still not straightforward. On the one hand, it seems that we can come up with some clear-cut cases: the green mould in the Petri dish is observable; sub-atomic particles are not. On the other hand, there are equally obvious grey areas: very large molecules or bugs on the borderline of microscopic, for example. Now this is not a problem for the realist. If her theory is appropriately justified and she takes it to be true, then no matter how we charac-terize the distinction between observable and unobservable, she will accept the objects put forward by the theory as 'out there' in the world. The anti-realist, of course, will adopt a different view.

Here's a different answer to our question 'What do theories tell us?':

A2: Theories tell us how the world is, in its observable aspect only (instrumentalism)

Realism has its problems, as we will see. In particular, unobservable entities have come and gone throughout history. So one option is to draw the above line and insist that the worth of theories lies not in whether they are true or false but simply in how useful they are when it comes to explaining and predicting phenomena. In other words, rather than telling us how the world is, theories should be regarded as nothing but instruments themselves that we use for predicting more observable phenomena (hence the name, 'instrumentalism'). This is a view that has fallen out of favour in recent years, mainly because theories function in scientific practice as more than mere instruments for prediction. That's why the best-known form of modern anti-realism adopts the following answer to our original question:

A3: Theories tell us how the world is, in its observable aspect and how the world could be, in its unobservable aspect (constructive empiricism)

This view accepts that theories play a role in science that goes beyond acting as simply prediction machines. However, for the sorts of reasons we shall look at below, it retains doubts about unobservable entities and processes and insists that whereas theories tell us how the world is when it comes to the observable features, we simply cannot be sure that it tells us how things are when it comes to the unobservable, only how the world might be.

Let us explore these positions in more detail.

Scientific realism

As we have just indicated, according to the 'realist', scientific theories correctly describe the way the world is: that is, scientific theories are:

- true,
- correctly describe what kinds of things there are in the world (observable and unobservable),
- correctly describe the ways in which these things are related.

Now this seems a straightforward position but even at this stage in the proceedings we need to exercise a little care. First, by 'truth' here, the realist means truth in the standard, no muss no fuss correspondence sense: that is, a statement is taken to be true if it corresponds to a state of affairs in the world. So, the statement 'electron has spin ½' is true if (and only if) the world is such that the electron does indeed have spin ½. Of course establishing that is another matter! But the core idea is that statements, hypotheses or theories in general are the 'truth-bearers' (they are the sorts of things that can be true or false) and the relevant states of affairs, such as the electron having spin ½, are called 'truth-makers' (at least in philosophical circles). The truth-makers, as the name suggests, make true the statements and hypotheses which then bear that truth.

However, it might seem too strong to say that (certain) theories *are* true, because we know from the history of science that theories come and go, that even those that are taken to be true at one time come to be abandoned and replaced at another. The obvious response for the realist to take is to acknowledge that earlier theories were not completely true but only approximately so, and that subsequent theories are improving on that approximation and taking us closer and closer to the truth. That seems a plausible picture but it turns out that filling in the details is more difficult than it might appear. There are other, more acute, problems that the realist has to face, however, as we'll see.

Second, we surely should not take *all* theories or hypotheses to describe the way the world is. What about speculative hypotheses – hypotheses that have passed a few tests but about which we still have doubts? That's a fair question and typically the realist restricts her realist attitude to *mature* theories: that is, those theories which:

● have been around for a while (that is, they're not speculative nor cutting edge),

● are generally accepted by the scientific community (there is a general consensus that they're on the right track),

● are seriously tested (they have survived falsification),

● are supported by a significant body of evidence (they have been verified).

These are the theories that tell us how the world is, at least as far as the realist is concerned.

Now, this might all sound plausible but can we give an argument for realism? You might think that a good argument is that many scientists are realists; indeed, such an attitude might seem to be a prerequisite for doing scientific research. After all, how can you investigate something if you don't think there's anything there to begin with? Well, first of all, not all scientists are realists. Many of the heroes of the quantum revolution, for example, concluded that it simply wasn't possible to give a realist interpretation of the new theory and retreated to a form of instrumentalism. Furthermore, even if adopting a realist attitude is necessary to do research (and it's not completely clear that it is – after all, I can believe that *something* is out there without accepting that the various aspects of my theory correspond to it), we could say that this is just a matter of psychology, of getting into the right frame of mind, rather than the basis of a convincing argument. Why should we, as philosophers of science attempting to understand scientific practice, adopt a realist attitude simply because scientists have to in order to do their work? Is there a better argument we can give? Indeed there is.

The 'ultimate' argument for realism (aka the 'no miracles' argument)

This is the argument that is most often given in order to try to convince someone to be a realist about scientific theories. It is nicely and famously summed up by the philosopher Hilary Putnam. He stated:

> The positive argument for realism is that it is the only philosophy that doesn't make the success of science a miracle.
> (H. Putnam, *Mathematics, Matter and Method,* Cambridge: Cambridge University Press, 1975, p. 73)

The central idea here is that realism is the best (perhaps even the only) explanation of the success of science. The main reason we're looking at scientific practice in the first place is because it is so massively successful: it has changed our world through its technological implications, giving us antibiotics, genetic manipulation, super-computers and iPhones, and it has changed our fundamental picture of the world, giving us evolution, curved space-time and quantum entanglement. More particularly, scientific theories are spectacularly successful in terms of making predictions that then turn out to be correct. How can we explain this? It's either an amazing (and repeated) miracle, or these theories have, somehow, got it right. Given our reluctance to accept miracles in this secular age – and this goes way beyond the odd burning bush – it would seem that the only conclusion we can draw is that the best explanation for the success of science is that our theories are true and tell us how the world is.

Furthermore, the realist might point out that this form of argument is no different from that used by scientists themselves with regard to their theories: just as scientists select a particular theory on the grounds that it's the best explanation of a phenomenon, so the realist argues that her philosophical view is the best explanation of the more general phenomenon of the success of science. So there's nothing odd or philosophically tricky about this argument – it's just the same kind of argument that scientists use. This forms part of a general view known as 'naturalism' which takes philosophy and science to form a seamless whole, and insists that philosophers should use the same sort of argumentative strategies.

So, the realists' ultimate argument for the truth of scientific realism is basically the same argument for the truth of scientific theories. That is:

- scientists argue that theory T is the best explanation of phenomena ∴ T is true,

- realists argue that realism is the best explanation of the success of science ∴ realism is true.

Now, we'll come back to this 'No Miracles Argument' shortly, but let's examine some of the problems this realist package faces.

Problem 1: The pessimistic meta-induction

The realist holds that our best, mature theories are true, or at least close to the truth. Enter the historian of science who laughs (evilly) and says 'been there, done that' and reminds us of all the theories through history that have been empirically successful but in fact are subsequently shown to be false, in the sense that they don't correctly describe what kinds of things there are in the world and/or don't correctly describe the ways in which these things are related. And if that was the case in the past, how can we be sure that our present, empirically successful theories won't also subsequently be shown to be false? And if that is the case, how can we be realists about these theories?

This argument against realism is known as the 'pessimistic meta-induction': it's a kind of inductive argument which uses examples from the history of science, rather than from science itself; so it's called a 'meta-induction' because it works at the level above that of science itself (the 'meta-' level); and it's pessimistic because it concludes that we can't regard our current theories as true and hence can't be realists. And it seems quite a powerful argument. What examples are typically given of past theories that were successful at the empirical level but we now agree are false? Here's a well-known list:

the crystalline spheres of Greek (Aristotelian) astronomy

the humours of mediaeval medicine

the effluvia of early theories of static electricity

catastrophist geology

phlogiston

caloric

heat as vibrations

vital force (physiology)

the electromagnetic ether

the optical ether

circular inertia

spontaneous generation.

Other theories can be found but these are some of the better known. So, here's the argument again:

> The history of science presents us with examples of successful theories that are now recognized as false;
>
> Therefore, our current successful theories are likely to turn out false;
>
> Therefore we have no grounds for adopting a realist attitude towards them.

Now, how can the realist respond to this argument? Well, she can point out that some of these examples were not particularly well developed, like the crystalline spheres or the humoral theory of medicine: that is, she can tighten up on the maturity constraint. In particular, she can insist that for a theory to be regarded as really mature, to be really worth of a realist attitude, it should make *novel predictions*: that is, predictions about phenomena that were not considered in the discovery or development of the theory in the first place. Recycling our iconic example from earlier chapters once again, the prediction that starlight would follow the curvature of space-time and be bent around the sun did not feature in the heuristic moves that lay behind, or in the subsequent development of, Einstein's General Theory of Relativity.

This extra criterion rules out some of the above examples; the crystalline spheres made no such predictions and nor did the humoral theory of medicine. But not all. Consider the caloric theory of heat, for example. This is the apparently plausible theory that heat is a kind of substance, called 'caloric', which flows like a liquid from a hot body to a colder one, and hence explains why hot and cold bodies brought into contact tend to reach the same temperature. This was an empirically successful theory that explained the expansion of air when heated (as caloric is absorbed by the air molecules) and also made novel predictions, to do with the speed of sound in air. But we now accept that the theory is false, and that heat is really the motion of molecules. So, if we had adopted a realist attitude towards the caloric theory we would have been caught out; it satisfies all the realist criteria but it was subsequently shown to be false. And if that could happen to the caloric theory, it could happen to our current well-regarded and accepted theories. Hence, we should not adopt a realist attitude towards them.

Problem 2: The underdetermination of theory by evidence

Which theories should we be realists about? Well, as we said, the ones that are empirically successful make novel predictions and are generally 'mature'. But what if we have two theories that are both equally empirically successful? Which one do we then take to be true? Consider two different theories about the extinction of the dinosaurs. Theory no. 1 suggests that it was due to a massive meteor strike that threw up huge quantities of dust into the atmosphere, blocking the sun, changing the climate and destroying ecosystems. Theory no. 2 posits that, on the contrary, it was due to massive volcanic activity that threw up huge amounts of dust into the atmosphere, blocking the sun etc. etc. Which one is true? The obvious response the realist can adopt would be to say that neither should be taken as true: both should be regarded as provisional hypotheses and we should withhold judgement until further evidence is obtained. So, when we discover evidence of a huge meteoric impact crater off the coast of Mexico, we can take that as further support for Theory no. 1 (see: https://en.wikipedia.org/wiki/Cretaceous–Paleogene_extinction_event).

But what if further evidence is found supporting Theory no. 2? What if we discover evidence of enormous lava flows in India, indicating dramatic volcanic activity around the time of the extinction? (See: http://science.nbcnews.com/_news/2013/03/28/17503180-volcano-eruption-theory-gains-backing-in-dinosaur-extinction?lite) What if, for each piece of evidence we find to support a given theory, we can find evidence supporting its competitors? The possibility of this is what is known as the 'underdetermination' of theory by the evidence: which theory we should accept as true is not determined by the evidence. And it forms the basis of another argument against adopting a realist attitude.

The central idea is as follows: for any theory T which is empirically successful and explains the phenomena, it is possible there could be an alternative theory T' which is just as empirically successful and explains the same phenomena but puts forward a different set of entities or presents a different way the world is. Now, how strong an argument the realist takes this to be depends on just how seriously she takes this as a possibility. Are there good cases of underdetermination in science? We've just seen how the underdetermination can be broken by new evidence that we discover. The realist might suggest that this will always be the case. But suppose the anti-realist is right: for whatever evidence we find for T, we can find further evidence supporting T'. Perhaps the realist can break the underdetermination by appealing to other factors.

For example, she might insist that we should believe whichever is the better explanation of the phenomena. This then raises the obvious question: What would count as a 'better' explanation? Now, we covered explanation in Chapter 4 but remember: both theories we are considering here are equally empirically successful, so in that sense they both 'account for' the phenomena. So, on the Deductive-Nomological view – according to which, we recall, a statement describing the phenomenon is deduced from the relevant laws plus background conditions – both theories would equally explain the extinction phenomenon, one in terms of astrophysical laws involving meteors (and laws of impact, meteorology etc., of course) and the other in terms of vulcanology (and also meteorology again and so on).

Likewise, adopting the causal account and appealing to the relevant cause isn't going to help – each theory offers a different cause of the extinction event and we have no way of manipulating or intervening with these causes. Even if we accept that the interventionist account shifts the focus away from human manipulation, considering 'what would have happened if things had been different' sorts of questions doesn't help us decide which of these two possibilities is the 'better' explanation and break the underdetermination.

Perhaps one explanation could be said to be more unified or more coherent than the other (and so we might shift to the unificationist account). So, the explanation of extinction events that cites volcanic action might require there to be more than one instance of such action having taken place, which might just seem less plausible than one big meteor strike, say. Now, the anti-realist can reply that appealing to plausibility seems pretty weak when we're supposed to be dealing with the truth, at least according to the realist. Perhaps it was just an unfortunate series of volcanic coincidences that led to the demise of the dinosaurs (and indeed, many think that the extinction of the dinosaurs can be explained by a combination of both volcanic action and a meteor strike).

Now, the realist can then counter by appealing to still other factors. Perhaps one theory is just simpler than the other, so the relevant laws given in the D-N type of explanation are easier to express and so the theory should be preferred on those grounds, since the laws offer a simpler and hence better explanation of the phenomenon. Of course, the realist then owes us an account of what simplicity amounts to – after all, Einstein's General Theory of Relativity doesn't seem that simple to most of us! But, it would seem that the realist can at least sketch out the basis of such an account: perhaps she might say that a theory which posits fewer unobservable entities in the world than another is simpler and to be preferred, so a theory which explains electrical phenomena in terms of one kind of charged object (negatively charged electrons) and their absence (positive), rather than two differently charged fluids, say (as in Benjamin Franklin's theory), is better.

However, the anti-realist can ask an apparently devastating question, which sidesteps the whole debate about what we mean by 'simple': What has simplicity to do with truth? Or, to put the point another way, why should the simpler theory be close to the truth? Unless the realist can link simplicity and truth in some way, trying to break the underdetermination by appealing to simplicity as a factor isn't going to help the realist's case. Now, the realist might just insist, as Einstein did, that the world just is simple, but insistence doesn't amount to an argument and such claims start to look like mere expressions of faith. After all, the universe could just be horribly complex, even at its most fundamental level, and it might turn out, then, that a very complicated theory is in fact closer to the truth. What the realist needs to show is that truth tracks or can always be associated with simplicity in some way, and so far, she hasn't been able to do that.

All is not lost, however. The realist has another card up her sleeve: she might say, look, this is all a bit too crude and in actual scientific practice we don't just consider the relationship between a theory and the evidence when we decide whether to accept it or not. We also consider other factors, such as the theory's coherence with other, already well-accepted theories, or with our background beliefs in general. So, consider our dinosaur example again. The theory that explains the extinction in terms of volcanic action gains extra support from the more general theory of continental drift. This explains a wide range of geological phenomena as due to the movement of huge 'tectonic plates' on which the continents sit. Where two such plates are moving apart, molten rock wells up from beneath the earth's crust and it was observed evidence that this was happening in the middle of the Atlantic, which provided conclusive support for the theory. Where these plates collide, one is forced underneath the other and the region where this happens suffers from earth-quakes and volcanoes. So, the presence of major volcanic action at the time of the extinction of the dinosaurs can itself be explained and made sense of in terms of the theory of tectonic movements, and evidence that there was such movement where and when the volcanic action occurred can be regarded as indirect evidence for this explanation of the extinction. That might provide further reason for preferring this hypothesis. In other words, establishing a relationship between this hypothesis and a broader set of background geological beliefs may help to break the underdetermination.

However, things are not that straightforward. The hypothesis of a devastating meteor strike also gains support from our recently acquired knowledge that such large interplanetary objects frequently pass close (in astronomical terms) to the earth. Indeed, it has been noted that dramatic extinction events appear to have taken place every twenty-six million years and it has been suggested that this is the period of time over which the earth encounters the 'Oort cloud', a large 'cloud' of rocks and debris left over from the formation of the solar system and from which asteroids and meteors periodi-cally emerge. Here we see astronomical background knowledge being appealed to in order to favour the meteor strike hypothesis. The problem is, as we now see, the adherents of each theory can appeal to different kinds of background knowledge in order to defend their claims and it may not yet be clear which set carries more weight.

Of course, the realist might pin her hopes on some further strengthening of the relationship between one of the underdetermined hypotheses and the relevant background knowledge, and appeal to that in order to break the underdetermination. But there is a straightforward response the anti-realist can make that appears to undermine the whole project: she can simply ask 'Why should we take the relevant background knowledge to be true?' Perhaps it too suffers from underdetermination with respect to the given body of evidence, so that some way had to be found to break that underdetermination too. But then if that involved further background knowledge – back-background knowledge – then the problem has just been pushed back a step. This is what the philosopher calls a regress and it's not clear where it stops.

We need to leave it there for the moment but at least we can see how this debate starts to open out into a whole range of other issues, to do with the relationship between theory and evidence, the role of factors such as simplicity and the impact of background knowledge on theory acceptance.

There is a final problem the realist must face, which goes to the heart of the motivations for this view.

Problem 3: The ultimate argument begs the question

We recall the 'ultimate' argument the realist gave for her position. This was that realism offers the best explanation for the phenomenal success of science – how else is that success to be accounted for, unless our theories are true or more generally have somehow got the world 'right'? And the realist insists that her argument for realism has the same form as the argument scientists themselves use for accepting one theory over another, namely that that theory offers the best explanation of the phenomenon. In other words, what the realist is doing is nothing tricky, philosophically speaking, but merely appealing to the same kind of argument – inference to the best explanation – that scientists use themselves.

But now the question arises: Do scientists actually use this form of argument? That is, do they conclude that a theory which is the best explanation of the phenomena is *true* and should be accepted as such? The answer is that some do, some don't, and in claiming that this is how all scientists operate the realist is guilty of assuming the very realist account of scientific practice that she is trying to defend. This nefarious practice is what philosophers call 'begging the question': you assume as part of your argument the very thing that you are trying to argue for! Clearly that's not going to stand up as a compelling argument, particularly if you're an anti-realist to begin with.

But then the anti-realist might legitimately be asked for her account of the success of science. In the next chapter, we will examine the best-known current form of anti-realism, but let's just note here what form that explanation will take. Basically, the anti-realist insists that we need to be careful in asserting that science is so tremendously successful. Clearly, *some* theories and technological spin-offs have been successful but to focus on this is to ignore the many other theories that were not so successful and which fell by the wayside. The success of current science only looks so impressive if we highlight the winners, and it appears less so if we bring all the losers into the picture: and of course, out of the vast array of theories put forward in the scientific journals and at conferences every year, only a very few will survive the wolves of experience; most will be falsified or shown to be incoherent. The anti-realist can make a nice comparison with the theory of evolution here: we notice that certain species appear to be fantastically successful in their particular ecological niches – the polar bear for example. One explanation is that there has to be something special about that species: that it was designed to be that way. Darwin offered a very different account in terms of 'natural selection' that eliminated the need for a designer – the species appears successful because its competitors were not 'fit' enough. Likewise, successful theories possess no special quality in terms of being true or whatever, they are simply the ones that are 'fitter' than their rivals, which could not survive scientific practice, red in tooth and claw! (But can you see where the metaphor cracks under the strain? We'll come back to this in the next chapter.)

9

Anti-realism

Introduction

In Chapter 8, we looked closely at the realist view of science. This takes the aim of science to be truth, not in some funny, post-modern sense, but in the sense of corresponding to states of affairs that are 'out there' – in the world. And the main, some say 'ultimate', argument for this view is that realism is the only position that doesn't make the success of science a miracle. This is the No Miracles Argument, or NMA. In other words, just as theories are accepted – the realist claims – because they are the best explanations of the phenomena they are concerned with, so realism is the best (indeed, the only) explanation of the success of science.

We then looked at the problems this position faces. First of all, the historically minded will say, 'Been there, done that, and didn't like this view in the first place,' pointing out that throughout the history of science apparently successful theories have come and gone: theories that, had the realist been around at the time, she would have accepted as the truth, or close to it, and which subsequently are thrown away as false, so why should we believe that our current theories, amazingly successful as they are, should be regarded as true, or approximately so? This is known as the 'Pessimistic Meta-Induction', or PMI.

Second, the situation may arise in which we have two theories that, it is claimed, are equally well supported by the evidence. This is the Underdetermination of Theory by Evidence, or UTE. If empirical success is supposed to be indicative of truth, how is the realist going to decide which theory is true, or closer to the truth? Now, the realist could always pin her hopes on further evidence coming in that will break the deadlock. But suppose that never happens? What could she appeal to then? She might suggest we go with the theory that is simplest, but then it's legitimate to ask what simplicity has to do with the truth? Or she could point to the way one of the theories is better integrated with our background knowledge than the other: but the same concern arises with regard to the background knowledge and so the problem is just pushed back a step.

And finally, there's the criticism that the NMA begs the very question at issue: that is, it assumes the very realist view it is designed to support. If you're not a realist, you're not going to accept the claim that scientists choose that theory which is the best explanation as true and so you're not going to be persuaded by the similar claim that realism is the best explanation of the success of science. Of course, the onus is now on you to come up with an alternative explanation of that success but, as we saw, that's not so difficult.

So, here's this chapter's fundamental question: How should we respond to these problems?

There are various answers to this question out there in the literature but here I'm going to focus on just three, well-known and, I hope, interesting alternatives.

Alternative 1: Constructive empiricism

This is perhaps the dominant form of anti-realism around in the philosophy of science today. Basically, it identifies the source of the PMI and UTE problems as being the appeal to unobservable entities and processes and urges us to restrict our belief to observable things only. Now, it is important to be clear how this form of anti-realism differs from earlier forms, such as 'instrumentalism'. The instrumentalist, as the name suggests, took theories to be nothing but instruments for the prediction of empirical phenomena and as such could not be regarded as true, or even approximately so. Theoretical statements – that is, statements about unobservable things like electrons, genes, the ego or whatever – are nothing but shorthand summaries of whole lists of observation statements. So, when a scientist states 'DNA is composed of a series of nitrogenous bases inter-connected by sugar and phosphate strands', the instrumentalist takes this to mean, 'When you do such and such an experiment, you will observe such and such a result.' (Obviously the list of observations in each case will be huge!)

The problem with this view is that it doesn't mesh with scientific practice. When a scientist says, 'All the evidence is in and it looks as if our theory is pretty close to the truth,' the instrumentalist has to translate that as, 'All the evidence is in and it looks as if our theory is a pretty good prediction device.' And when scientists talk about electrons, genes, the ego, whatever, the instrumentalist has to say, 'Ah, what you're actually talking about are mammoth lists of observations': to which, the scientists might well respond, 'No, what we're talking about are electrons, genes, the ego or whatever!' On this view, we can't take the language of science literally and we have to translate all the talk and beliefs of scientists in terms of observations.

The constructive empiricist, on the other hand, takes the language of science literally. She agrees that when scientists talk about unobservable entities, their talk is, indeed, about these entities and is not mere shorthand for long lists of observation statements. And she also agrees that theories are the kinds of things that can be true. However – and here's the twist – the constructive empiricist adds what she takes to be a healthy dose of scepticism to the pot. How do we know that theories are true, she asks? In particular, how do we know that theoretical statements refer to the unobservable entities they purport to? If we buy into the empiricist premise that all knowledge is only of the empirical, that is, the *observable* (what we can observe with the naked eye), then we clearly cannot *know* whether electrons, genes or the ego exist, nor can we know, therefore, whether theories are true or not. They might be, we just can't know.

On this view, then, we shouldn't believe theories to be true, or approximately so. What attitude should we take towards them? Well, what scientists do, as we have seen, is test their theories, seek empirical support for them and try to determine whether they are adequate in terms of accommodating the relevant observations. So, rather than believe theories as true, we should simply *accept* them as empirically adequate. As far as the constructive empiricist is concerned, this is the appropriate attitude we should adopt towards theories and, furthermore, we should drop the realist view that science aims at the truth and acknowledge that its aim is *empirical adequacy*. Here's what the 'founder' of constructive empiricism says:

> Science aims to give us theories which are empirically adequate; and acceptance of a theory involves as belief only that it is empirically adequate. ... a theory is empirically adequate exactly if what it says about the observable things and events in this world are true – exactly if it 'saves the phenomena'.
>
> (B. van Fraassen, *The Scientific Image*, Oxford: Oxford University Press, 1980, p. 12)

What do theories tell us, then? On the realist view, theories tell us how the world is. But according to the constructive empiricist, we can never *know* how the world is, since we can never know its unobservable aspects. On this view, theories tell us how the world *could* be: that is, they provide us with useful stories of what the world might be like, but we can never know if these stories are actually true or not.

Now, you might find this a beguiling position; or you might think it's clearly mad! Before we make any critical judgements, let's see how it handles the three problems above.

So, first of all, how does constructive empiricism overcome the PMI problem? We recall that at the core of PMI lies the claim that the history of science presents us with case after case of radical change at the level of unobservable entities. Phlogiston, caloric, the ether have all been proposed by the respective theories and these theories have even enjoyed some empirical success, but they were all abandoned and these entities dismissed as unreal. Nevertheless, there is steady cumulative growth through the years at the level of the observable consequences of our theories. Of course, sometimes what initially appear to be good experiments are discovered to be flawed or problematic, but leaving these cases aside, the history of science does seem to present us with an accumulation of empirical results. (Some philosophers and sociologists have disputed this but we'll come to that in the next chapter.)

Now, it would seem that constructive empiricism can easily accommodate this. If we think of a theory as merely telling us how the world could be, then we shouldn't be surprised, or at all bothered, when the evidence tells us that no, it couldn't be like *that*. Of course, that doesn't provide conclusive evidence that the world *is* how the next theory proposes it is: again, this is just another way it could be. So the radical changes at the level of unobservables are nothing more than changes of story, from 'The world could be like this' to 'Or it could be like that ...' In every case, we can't know for sure. And as the level of evidence accumulates, each successive theory can be viewed as more empirically adequate than its predecessor and so the growth of empirical knowledge can be accommodated.

Secondly, how does constructive empiricism overcome the UTE problem? This is even less of an issue. We recall that UTE states that there may arise situations in which we have two theories both equally supported by the evidence and hence we cannot believe either theory to be true. Indeed, says the constructive empiricist, nor should we! Nevertheless, we can accept both theories as empirically adequate. Of course, as a practising scientist, you may have to choose to work on one rather than the other simply because you don't have enough funding or expertise to work on both. Or you may decide that one is simpler, or easier to work with, than the other. That's all fine; the reasons for your choice have nothing to do with the truth of either theory. The choice to work with one or the other will be made on purely 'pragmatic' grounds.

Finally, what about explaining the success of science? Obviously, the constructive empiricist will not go with the NMA. Instead, she might question the sense in which, without the NMA, the success of science would be a miracle. As we noted at the end of the last chapter, this success appears striking but perhaps that is only because we focus on the successful theories and forget about all the others that fell by the wayside. Take a stroll through the university library sometime, run your fingers along the bound and collected volumes of the *Journal of Neurophysiology* or the *Physical Review* or the *Journal of Chemical Ecology* or any one of the many specialist scientific journals and pull out one of the dusty volumes from the early years. Just look at all the theories and hypotheses that were proposed but have subsequently been abandoned. Many of them, of course, were immature, half-baked even, but given such a plethora, is it any wonder that sometimes, some of them get it right?

Compare this, again, with the situation in biology: through mutations or recombinations the DNA of an organism changes. These changes can be beneficial, harmful or neutral. If a change is harmful, given a particular local environment, then it may be unlikely that the offspring that inherit that mutation will themselves survive to reproduce and so the mutation dies out. If it is beneficial, again in the context of a particular environment, then the mutation may confer some advantage on the organisms that inherit it and hence the mutation spreads. As this process continues, entirely new species will form and we end up with the fox, which can live just about anywhere from the tundra to urban parks and can eat just about anything and we think, 'Wow, this species is amazingly successful': but it only seems miraculous if we forget about all the evolutionary false starts due to harmful mutations along the way. The actual explanation is rather prosaic: there were lots of changes, only a few of which were beneficial and those are the ones we notice. It's the same with scientific theories: we tend to forget all the false starts and falsified hypotheses and by isolating the really successful ones treat that success as something in need of a realist explanation.

So, let's sum up the core ideas of constructive empiricism:

1 We have knowledge only of the *observable* (this is the empiricist feature), where what is observable is what could be observed by the naked eye in principle, that is as described and understood by science itself (so, for example, Jupiter's moons are observable, because science tells us that we could travel out beyond the asteroid belt and observe them with the naked eye; electrons are not observable because, despite certain cheesy science fiction movies, science itself tells us that we could never shrink down to see them with our own eyes);

2 Unobservable entities and processes may exist but we can never *know*;

3 Theories may be true but we can never *know*;

4 Theories may nevertheless be accepted as empirically adequate; and

5 Empirical adequacy, not truth, is the aim of science.

Now hopefully I've convinced you that this is an interesting view and, more than that, a viable alternative to realism. However, it too faces problems.

First of all, as we have just emphasized, it is grounded on the idea that we can have knowledge only of that which is observable with the naked eye. Now you might object that this relies on making a clear distinction between the observable and unobservable, one that we may have real trouble actually making. In times past, philosophers tried to draw the distinction, in linguistic terms, between observation statements and theoretical ones, but gave it up as hopeless. The modern constructive empiricist doesn't think the distinction can be drawn that way, but rather in terms of the entities themselves. So, we're observable, and so are Jupiter's moons, but electrons are not. In between we might encounter a grey area where it's just not clear whether the entity concerned – very large molecules perhaps, or very small bugs – counts as observable. But that just means that 'observable' is a vague term (like 'bald') and as long as we have a good idea of when it can be employed and when not, there shouldn't be any problems.

More significantly, perhaps, you might feel that taking 'observable' to mean 'observable with the naked eye' is just way too restrictive. What about the use of instruments like the microscope? Don't scientists talk of 'observing' things through such instruments? Indeed, they even talk of 'directly observing' the core of the sun using highly specialized detectors that record the flux of neutrinos that we considered in Chapter 5. However, this is where the constructive empiricist reminds us of that second term in her name – she is an empiricist and that means taking a particular stance with regard to what counts as knowledge, one that emphasizes the role of experience, whether that be understood in terms of sensory data, or is extended to include the connections between these data. To insist that you have a broader understanding of 'experience' is just to adopt a different stance and if the constructive empiricist can account for everything you can, in particular if she can account for scientific practice, and, moreover, avoid the PMI and UTE problems, then it's not clear on what grounds you can say that your stance is better!

But what if we were to perform some Frankenstein type experiment and replace someone's eyes with twin electron microscopes? Such a person could presumably claim to 'observe' bacteria, the crystalline structure of various surfaces, even clusters of atoms (the 'Size and Scale' website even categorizes this under 'New Sets of Eyes'). Or imagine that the SETI project finally pays off and we find ourselves in contact with an alien species whose eyes have evolved differently so they can see to a level that we can't (just as birds, for example, can see polarized light that we can't). Doesn't that suggest that the constructive empiricist's distinction between what is observable and what is not is somewhat arbitrary?

The constructive empiricist answers as follows. We need to be clear that when we look at the beautiful images produced by a scanning tunnelling microscope (see, for example, http://nobelprize. org/educational_games/physics/microscopes/scanning/index.html), we are looking at images that have been produced by a physical process very, very different from the impact of light on the human eye, one that involves the 'tunnelling' of electrons between the surface of the body and a tip so sharp it consists of just a single atom, producing an electrical signal that is kept constant by raising and lowering this tip, and this raising and lowering is then recorded and enhanced by computer to yield an image. So our Frankenstein operation would have to involve a bit more work than simply plucking out someone's eyes and replacing them with an instrument like this and any alien that 'sees' via such a process would have to have a very different physiology from ours. Indeed, the constructive empiricist would insist, it would be so different that we would have to conclude that such aliens, or the results of our monstrous experiment, could not be counted as a part of our 'knowledge community', in the sense that what counts as knowledge for them would have to be very different from what it is for us.

The second major problem has to do with the explanation for the success of science. We saw that this provides the motivation for realism, through the NMA. Now, is the Darwinian explanation of the success of science offered by the constructive empiricist itself adequate? Let's return to the analogy with evolution and 'survival of the fittest'. The underlying notion of the 'fitness' of a species is now understood in genetic terms, as we indicated in our sketch above. What would correspond to such terms for a *theory*? Let me put it another way: we now understand how a particular species, like the European fox, for example, is so successful in terms of the interaction between, ultimately, genetic changes and the particular environment the foxes' evolutionary ancestors found themselves in. The 'environment' for a theory might be taken as the empirical world, with experimental results leading to the extinction of certain theories and allowing the survival of others. But what would count as the underlying mechanism, analogous to the genetic make-up of an organism, that drives the changes in theories? It's hard to see that there could be any such mechanism and so the analogy begins to look a bit ropey. The realist, of course, has an answer: a given theory is successful in that particular empirical environment because it has, in a sense, latched on to the world; it has 'got' the world right.

Alternative 2: Entity realism

Even if you agree that constructive empiricism is too sceptical a stance and too restrictive in what counts as knowledge, you might be reluctant to return to full-blown realism. Is there no more modest form of realism that satisfies the feeling that we can know how the world is, in both observable and unobservable terms, and solves the problems faced by its bloated cousin? Here's another alternative that might do the trick.

First of all, let's recall that the source of the PMI problem is the apparent abandonment of certain unobservable entities throughout history; and the source of the UTE problem is the focus on the truth of theories. The view known as 'Entity Realism' (a view developed by Hacking and nicely described in his book, *Representing and Intervening*, already much cited here!) offers a way through these difficulties by urging us to tear our philosophical attention away from theories and the thorny issue of whether they can be held as true or not, or might be true only we can never know it, or whatever, and instead focus on those unobservable entities that we are confident exist, not because they are presupposed by some theory, but because we *use* them. It's this pragmatic feature of entity realism that marks it out from other positions in the realism–anti-realism debate. The core ideas, then, are the following:

1 Some entities are *retained* through scientific change: e.g. the electron, the gene;

2 Our belief that these entities exist has nothing to do with the *truth* of theories, but with their *practical manipulation* in the creation of phenomena.

Here's what Hacking says:

> Experimental physics provides the strongest evidence for scientific realism. Entities that in principle cannot be observed are regularly manipulated to produce new phenomena and to investigate other aspects of nature. They are tools, instruments not for thinking but for doing. ...
>
> The experimentalist does not believe in electrons because ... they 'save the phenomena'. On the contrary, we believe in them because we use them to create new phenomena.
>
> (I. Hacking, 'Experimentation and Scientific Realism', *Philosophical Topics* 13: 154–72; see also I. Hacking, *Representing and Intervening*, Cambridge: Cambridge University Press, 1983)

He gives the example of scientists spraying a stream of electrons at a tiny, niobium ball in order to change its charge in an experiment to detect the presence of sub-nuclear particles called quarks. We don't need to bother with the details of the experiment; what is important, is the fact that the electrons are regarded as nothing more than just a tool which the scientists manipulate to create a new phenomenon. This has given rise to a famous slogan summarizing Hacking's view: if you can spray 'em, they're real! The electrons are just something that can be more or less taken down off the shelf and used to achieve the desired effect. And just as the car mechanic doesn't fret over whether her wrench is real or not (at least not unless she's taken some philosophy classes), so the scientist doesn't and shouldn't worry about the reality of electrons and other unobservables.

Now, how does this position overcome the PMI problem? We recall that the point of the PMI argument was to cut the realist's link between the empirical success of theories and belief in the existence of the entities posited by those theories. However, the entity realist maintains that the belief should not be based on the idea that the theory is true, or that it has latched onto how the world is, but that it should follow from the fact that we can use the entities concerned. Many of the entities that have come and gone in the history of science fail this test: they could not be used to create new phenomena and so were rightly rejected. Nevertheless, some entities can be used in this way and thus we have grounds for *optimism*, but these grounds have nothing to do with the success of the associated theories; rather they have to do with the use to which these entities are put. Consider the humble electron again: the associated theories have changed quite radically, from theories which took electrons to obey the classical mechanics of Newton, to the new quantum theory which suggested they had a wave-like aspect, to quantum electrodynamics which presents them as simply bumps in a quantum field, to today's string theories and so on. Despite all these changes, scientists have continued to believe in the existence of electrons because they have become an indispensable tool.

So, how does entity realism overcome the UTE problem? This is even easier to tackle. We recall that UTE insists that of two theories supported by the same evidence, we cannot believe either theory to be true, hence we cannot believe in either of the entities posited by the theories. Well, as we've seen, entity realism argues that belief in the existence of certain entities has nothing to do with belief in the truth of the associated theories. Indeed, Hacking maintains that scientists typically use different or even incompatible models of the electron, for example, without worrying about the *truth*. We can still believe that such entities *exist* even when faced with UTE type situations.

Finally, then, how does this view explain the success of science? Remember, for the realist this is really important. She employs the same argumentative strategy as scientists themselves, or so she claims: namely, we take the best explanation of the phenomenon at hand and regard that as the truth. In the case of science, the phenomenon might be the bending of starlight around the sun, for example, and the theory would be Einstein's General Theory of Relativity; in the case of the philosophy of science, the 'phenomenon' is the success of science itself and the 'theory' is realism. However, the entity realist just isn't interested in the supposed truth of theories, as this is not indicative for what we should take as 'real'. As we've already said, the empirical success of theories can be deceptive, leading scientists to accept the existence of entities subsequently shown not to exist. The entity realist has a different view of success: science should be deemed successful not because it allows us to better represent the world, and say how the world is, but because it allows us to *intervene* in the world, by, for example, creating new phenomena and new technologies. It is intervention and not representation that we should be focusing on and it is the fact that we can use them as tools for intervention that leads us to believe in electrons and other unobservable entities.

Now, this is a powerful and quite compelling view but it too faces certain problems.

First of all, it has an obviously unpalatable consequence: what if you're faced with an entity, or rather with a hypothesis positing an entity, which you can't manipulate and can't use to intervene in the world? The entity realist would presumably have to insist that you have no good grounds for taking this entity to exist. Now, stepping outside of the domain of physics for a moment, that may not present much of a problem for the chemist, say, since she can argue that as she uses certain kinds of molecules to produce certain effects and create certain kinds of phenomena, she can claim these molecules exist. Likewise, the biologist who uses certain enzymes to snip strands of RNA into pieces in order to create certain genetic phenomena has grounds for regarding these enzymes, at least, at real. But what about the psychologist who talks about the ego, say? They seem to be on much thinner ice. Perhaps that's a good thing; perhaps this is a way of winnowing out all the 'dodgy' entities and leaving only those we really should take as real (really real!).

But even in physics, or rather astrophysics, there may be problems. Astrophysicists have noticed a kind of phenomenon whereby very similar objects appear to be symmetrically reproduced across certain regions of space. Consider, for example, the 'Einstein Cross', where a 'cloverleaf' of four bright spots can be seen at the centre of a distant galaxy (see http://apod.nasa.gov/apod/ap001010.html). Now, most galaxies have only one nucleus, so this is an odd phenomenon and astronomers have tried to explain it by suggesting that in fact what we're seeing is the light from a far distant object known as a 'quasar', which is bent and split by the gravitational field of the intervening galaxy to produce four images. The galaxy is acting as a 'gravitational lens' (another beautiful example has been photographed by the Hubble telescope: see http://hubblesite.org/newscenter/newsdesk/archive/releases/1996/10/).

Now, many astronomers have come to accept the existence of these gravitational lenses because they explain a number of otherwise bizarre phenomena. And it's easy to see how the explanation goes: here's something really odd – four bright spots at the heart of a galaxy, for example. The chances that this is just a galaxy with a very unusual heart are really low; a better explanation – indeed, the best one – is that we're seeing another gravitational effect: the mass of the galaxy is so huge that it distorts the surrounding space-time sufficiently to refract and bend the light of a distant object, creating the four images. However, as far as the entity realist is concerned, this is not good enough; we cannot believe in the existence of gravitational lenses until we can use them and manipulate them to produce new phenomena. Now what are the chances that we're going to be able to use the centre of a galaxy, the way a mechanic uses a wrench, any time in the future?! This puts the entity realist out of step with the top scientists in astrophysics but perhaps that's a bullet she's prepared to bite.

That's not the only objection, however. The entity realist accepts that electrons, genes, etc. (but not gravitational lenses, or black holes) exist; but what are they? If we say an electron is a charged sub-atomic particle, or a bump in a quantum field or the vibrating end of a quantum 'string' or whatever, where have we got that description from? A theory, of course. But how can we say what an electron, gene, whatever, *is*, if our theories about it change or if we have incompatible theories about it? As we've already noted, our description of the electron has shifted quite dramatically in the past hundred years or so, from being a small chunk of matter, to a wave-particle, to a bump in the Great Quantum Field, to the manifestation of a multi-dimensional superstring, to ... But if we focus on these descriptions, we're faced with something like the return of PMI! The entity realist might be able to say that electrons exist, because she can use them like tools, but she can't say with any confidence what they are, because history teaches us that our current description may soon go the way of those held ten, fifty, a hundred years ago. However, if we can't say what the electron is, isn't our belief that it exists empty?

Again, the entity realist may just have to swallow that and agree that all she can say is that there's *something* out there, it is charged in such and such a way, it has a certain mass and so on. But here's a final objection that many people take to be a real hurdle that has to be overcome.

The entity realist, as we have seen, focuses on the use scientists make of certain entities. But are electrons, enzymes and the like really on a par with a mechanic's wrench? After all, you can't actually pull down a big box of electrons off the shelf and start throwing them around. What scientists actually do is use an electron gun, which produces an appropriately focused beam of electrons which can then be targeted on a niobium ball, or the inside of a TV screen or computer monitor. It's the electron gun that is more like a wrench, to be used by the scientists to achieve the effect they want. Now, the electrons are unobservable – that's the crucial issue, of course – so what the construction and use of the electron gun rely on is an understanding of certain properties of the electrons (such as charge and mass) and the laws they obey. These laws may not be super-high level and abstract; they may be cobbled together in such a way that they apply only to the particular situations in which the electrons are produced, but it is these that the scientists rely on. In other words, the scientists must accept these low-level laws as true in order to achieve the effects they want. So, when we use electrons, say, to create new phenomena, we're relying on the truth of 'low-level' (causal) theories about electron behaviour. And these low-level laws and theories are accepted as true because they are empirically successful. But if we focus on these low-level laws, we're faced with something like the return of UTE! Suddenly entity realism doesn't look that different from the more standard form.

These sorts of objections haven't ruled the position out and many philosophers of science continue to develop it, particularly those who feel that the analysis of science tends to be too theory oriented and needs to focus more on pragmatic and experimental matters. However, there is another form of realism that goes to the other extreme and embraces the theoretical. Let's take a look at that before we move on.

Alternative 3: Structural realism

Let's go back to the PMI and look a little more closely at the history of science. And let's take another example: the history of light. Newton famously thought light was composed of tiny particles that underwent 'fits' when they passed from, say, air to water, leading to the phenomenon of refraction, which contributes to the formation of rainbows as discussed in Chapter 4. Then Young proposed that light is in fact a wave and Fresnel developed this theory further, producing a set of equations (now known as, surprise, surprise, the Fresnel Equations) which describe the behaviour of light when it passes from one medium – air, for example – to another – glass, say. When a critic pointed out that if light really was a wave, under the right conditions we should see a white dot in the shadow cast by an illuminated disc (due to diffraction around the edges of the disc), Fresnel ran the experiment and was as surprised as the critic when a white spot was observed. Maxwell brought light under the umbrella of his theory of electromagnetism (remember Hertz's experiments?) according to which it was conceived as an oscillating electromagnetic wave. Then came quantum theory and Einstein (again) argued that light had to be seen(!) as possessing particle-like qualities, so that it demonstrated the famous quantum wave-particle duality. Subsequently it too was regarded as a kind of quantum field and so the story of its changing nature continues.

Now this looks like nice grist to the PMI mill: light as a Newtonian particle has been abandoned, as has light as a wave, so we have no good reason to suppose that in future years the idea of light as a quantum field will not also be consigned to the dustbin of history. But perhaps this is too hasty. Perhaps there is something that is retained throughout these dramatic theoretical shifts, something more than just all the empirical evidence that the constructive empiricist focuses on. After all, we still use Maxwell's equations (in certain circumstances) in the quantum era, and even after Maxwell proposed his theory, scientists still used Fresnel's equations. Indeed, they drop out of or, more precisely, can be deduced from Maxwell's theory if certain conditions are applied and in this sense they are retained despite all the shifts in our views of what light actually *is*. These equations can be understood as representing the underlying structure of reality and the view that *structure* is what is retained through theory change and what we should be realists about is known as *structural realism*. Its core ideas are as follows:

1   *Structure* is retained through scientific change;

2   *Structure* is what we should be realists about.

This is actually quite an old idea and if you look back through the last hundred years or so of commentaries on science you find it cropping up again and again. Poincaré, for example, was a famous (and brilliant) mathematician and physicist (he came within a whisker of discovering the Special Theory of Relativity, for example), who also thought deeply about the nature of science. He also noted that certain equations are typically retained through theory change and wrote,

> ... if the equations remain true, it is because the relations preserve their reality. They teach us now, as they did then, that there is such and such a relation between this thing and that; only the something which we then called motion, we now call electric current. But these are merely names of the images we substituted for the real objects which Nature will hide for ever from our eyes. The true relations between these real objects are the only reality we can attain.
> (H. Poincaré, *Science and Hypothesis*, New York: Dover, 1905, p. 162)

The idea then, is that all that we can know about reality is captured by the equations representing the relations between things, whose true 'natures' we can never really know (so to that extent the PMI is right).

How does this view overcome the PMI problem? Well, the answer should be obvious: PMI insists there is radical change at the level of unobservable entities; but it overlooks the fact that there is also retention of certain *structures* at this level. Shifting our attention away from the entities and on to the structures, it is the latter we should be realists about.

How does this view overcome the UTE problem? This is a little trickier, but one response runs as follows: UTE is supposed to lead us to conclude that we cannot believe either theory to be true, but that's OK, because the structural realist doesn't take the entire theory to be true, just those structural aspects that are retained through theory change. So, the structural realist will insist that in order for both theories to be empirically successful, they're going to have to possess certain equations or structures in common and it is that common part we should believe to be true (or approximately so). Now, if the anti-realist can come up with examples of UTE where there are no common (structural) parts beyond the empirical level, then the structural realist will be scuppered. The extinction of the dinosaurs could be one example, although the structural realist will follow the ordinary realist in arguing that further evidence will surely settle that case one way or the other.

Finally, how does this view explain the success of science? Here the structural realist typically follows his non-structural cousin and argues that the success of science gives us good reason to suppose that our theories correctly describe the world, at least with regard to its structural aspects. In this sense the structural realist wants to present herself as less radical than the constructive empiricist and not as restrictive in her beliefs as the entity realist.

Now, as we noted above, this form of structural realism holds that all we can *know* is the structure of the world and we just have to remain agnostic about the nature of the entities. There is another strand, however, which insists that it is not that all we know is structure, but all there *is* is structure. The motivation for this is quantum physics. A proponent of the above form of structural realism wrote that,

[t]he structural realist simply asserts ... that in view of the theory's enormous empirical success, the structure of the universe is (probably) something like quantum mechanical.

(J. Worrall, 'Structural Realism: The Best of Both Worlds?', in D. Papineau (ed.), *The Philosophy of Science*, Oxford: Oxford University Press, 1996, pp. 139–65 (originally published in *Dialectica* 43 (1989): 99–124, 163)

But according to quantum physics, the 'nature' of the entities of the world as objects is deeply problematic. This is something the original heroes of the quantum revolution spotted and they suggested that, according to the theory, the fundamental entities should not be regarded as individual objects, in the way that tables, chairs and people can. That's probably enough to make you wonder about the nature of these entities. It turns out, however, that the theory *is* consistent with the framework of individual objects, if understood a certain way. So now it appears we have another kind of fundamental underdetermination, only this time with the theory supporting two very different basic interpretations: in one the entities of the theory are not individual objects, in some sense; in the other they are. Anti-realists such as the constructive empiricist allege that this raises yet another problem for the 'standard' realist, since if she can't even say whether the objects she believes exist are individuals or not, what good is her realism?

This second form of structural realism responds to this challenge by suggesting that we should drop the notion of object from our theory altogether, so that what theories are about, on this view, are nothing but structures, pure and simple (see J. Ladyman and D. Ross, *Everything Must Go*, Oxford: Oxford University Press, 2007). Well, perhaps not so simple as it is not entirely clear what it could mean to say that the world is fundamentally just structure. A common understanding of a structure is that it consists of a family of relations holding over a set of objects. So, consider the genea-logical structure of your family, with relations such as 'father of' and 'daughter of' holding between various people. But if the objects are removed from the picture, what do the relations hold between? And how can relations hold without any relata? These are crucial questions but to go any further would take us way over the cutting edge and into the abyss! All I can say at this point is that explicating this form of structural realism is the top priority for this kind of realist (like me! And if you're really interested, you can see where my research has taken me in *The Structure of the World*, Oxford: Oxford University Press, 2014).

As with the other positions, structural realism also faces problems. First, in its focus on mathematical equations, this position seems to be oriented towards the more mathematical sciences, such as physics. What about biology, or even psychology, where there is far less mathematization? Can structural realism find a place in these fields too? One answer is a blunt 'yes', since the notion of structure is broad enough that one can argue that maths is just one way of representing it. However, there is a lot more work to be done in developing structural realism in a biological context, say.

A second problem is associated with the question of whether it is always the case that structure is retained through theory change. What if the structures themselves change? If that happens, then we've lost one of the main advantages of going structural, which is to respond to the PMI. However, even granted that there has to be some change for science to progress, it is not clear that the structures the realist is interested in change so radically that structural realism is fatally undermined. Finally, however, doesn't the above response to the UTE problem *assume* precisely what the realist needs to *show*? It merely expresses the hope that in any such cases, there will always be common structure. But what if two such empirically equivalent theories don't have any structure in common? Then we would see the return of UTE as well. As in the previous problem, what we need to see are some concrete examples, and these have not been forthcoming, at least not so far.

Conclusion

There is a variety of views on offer. The ones I've covered here – standard realism, constructive empiricism, entity realism and structural realism – are just some of the better-known. Which one you think is the 'best' account will depend not only on your understanding of scientific practice, its aims and its history but also on your philosophical views about what we can know. Any argument in favour of one position runs the risk of 'begging the question' against the others. All I've tried to do here is sketch the main arguments for and against, and bring you up close to the 'cutting edge' in this area. Now we're going to consider a broader form of anti-realism, one that gets its force from the suggestion that scientific practice, and in particular scientific change and progress, is not driven by observations and empirical support, but by social, political or economic factors.

Exercise 4

Based on Chapters 8 and 9, here's a starter just to get those brain juices flowing:

Q1 What is the relation between truth, scientific theories and scientific confirmation from the perspective of the realism/anti-realism debate?

Now, read the following blog entry: http://welovephilosophy.com/tag/scientific-realism/

Q2 What is meant by 'common-sense realism'? What are its 'two facets'?

Q3 How does the scientific realist 'come to grips' with the uncertainty generated by the history of science? Do you think this is a good response to the Pessimistic Meta-Induction?

Q4 What is the indispensability argument for mathematical realism? What is the main problem faced by this form of realism? Does common-sense or scientific realism face a similar problem?

Q5 What is moral realism? Why is it harder to defend than common-sense or scientific realism?

Q6 Is the 'constructive empiricist' a common-sense realist?

Q7 What kind of realist (if any!) are you?

And here are the more advanced posers:

A What is the main difference between instrumentalism and constructive empiricism?

B Is the distinction between the observable and the

unobservable (as drawn in constructive empiricism)
problematic?

C   Is van Fraassen right in claiming that constructive empiricism
is equally as rational as scientific realism?

D   What does constructive empiricism say about the idea that
scientists often choose between theoretical alternatives
on the basis of their different explanatory virtues? What
does the realist say?

10

Independence

Introduction

Well, you have your theory, it explains various phenomena and has had some empirical success, and on that basis you believe it tells you how the world is, if you're a realist, or how the world could be if you're a constructive empiricist. But up pops a sociologist and points out that you're a child of your time, the product of specific socio-economic and political conditions and therefore so is your theory. It says less about how the world is or could be and more about those conditions. Now that's a strong line to take but as we'll see, it has some force. In effect, the sociologist is raising the following fundamental question: Is science independent of its social context?

One answer is: of course not! There is clearly a sense in which the socio-economic and political conditions have to be right for science to flourish. After all, if there isn't appropriate funding, whether from universities, government or private businesses, or appropriate institutional structures, which can support the right training and career development, then at the very least science will not have the support it needs. We can even look back at the history again and suggest that the scientific revolution of the seventeenth century would not have happened without the shift from a feudal system, or the great developments of the nineteenth century would not have taken place without the industrial revolution. We can also try to answer the question why the scientific revolution occurred in Western Europe rather than, say, China by focusing on these specific socio-economic conditions. But interesting as these suggestions might be, this answer is basically trivial in that it offers no threat to the *objectivity* of science: the conditions might have to be right for science to flourish but they don't determine the content of scientific theories in the way our sociologist friend seems to suggest.

Another answer is: of course not! These socio-economic and political conditions are reflected in the actual content of the theories, in various ways, perhaps quite subtly. Consider Darwin's theory of evolution, for example, with its emphasis on survival of the fittest. Is this anything more than a reflection of the prevailing Victorian ethos, according to which the 'fittest' happen to be white, British males? This answer is highly non-trivial of course, and undermines the objectivity of science, or at least replaces that notion with a very different one.

By 'objectivity' here is meant something like the following (at least in part): science is value-neutral in the sense that 'contextual' values (preferences, beliefs, interests, etc.) are subjective values of an individual or the cultural biases of an entire society that have no place in scientific theories or *should* have no place in scientific theories. Here then is this chapter's fundamental question: How might social factors affect science?

Science as a social activity

As we've already indicated, there are clearly senses in which science can be considered a social activity but which do not undermine its objectivity. Here are some of those senses:

1   **Social factors may determine *what* science investigates**
    With limited funding, not every problem, interesting phenomenon or significant medical condition can be investigated. Here's an example that generated huge debate, not just among lay-people but also among scientists themselves: in the 1980s it was decided to construct a huge particle accelerator in Texas, big enough and powerful enough to reach energies sufficient to reveal one of the Holy Grails of particle physics, the Higgs boson, mentioned in Chapter 4 as the particle that gives other particles their mass. However, by 1993, costs had spiralled to $12 billion,

almost three times the original estimate and equivalent
to NASA's entire contribution to the International Space
Station. Other scientists, including physicists, began to raise
concerns about the funnelling of federal funds away from
other areas of research. Political considerations also came
into play as Democrat institutions at both state and federal
level questioned whether they should be supporting a project
begun under the Republicans and at such cost. Eventually
the project was cancelled, having spent $2 billion and leaving
some very large tunnels under the Texas countryside. As
for the 'God particle', as I mentioned earlier, it was finally
observed in 2012 at the Large Hadron Collider, at CERN, in
Europe (see: http://home.web.cern.ch/topics/higgs-boson).

Shifting to medicine and health care, the allocation
of resources in these areas has long been a source of
controversy. A useful statement of the difficulties involved
in funding fundamental research in the current challenging
financial climate has been given by the Director of the US
National Institutes for Health at http://www.nature.com/news/
nih-chief-keeps-hopes-afloat-1.15723. At some point social,
political and sheer economic considerations come into play,
leading to the kinds of inequalities that spark concerns among
activists, patients and health care professionals themselves.
We'll be looking at examples of these in the next chapter.

Now, do these kinds of considerations undermine the
objectivity of science? No: this is just a matter of allocation of
resources.

2   **Social factors may determine *how* science investigates**
There are different ways one might go about doing science:
different ways of conducting experiments for example. Some
of these might be deemed socially or ethically unacceptable
and in this manner social conditions may influence scientific
practice. So, for example, scientific research involving
human subjects is generally subject to quite rigorous
ethical standards, which may rule out certain experiments,
no matter how scientifically interesting. However, other
societies with lower or different standards may have no such

compunction. So, Nazi scientists, for example, conducted appalling experiments on concentration camp inmates, subjecting them to horrific extremes of temperature (by immersing them in ice-cold water, for example) in order to 'test' the resilience of the human body. Clearly we would deem such experiments as utterly unacceptable and would refuse to condone them. But what about the results of the Nazi experiments themselves? Should they be used to help design survival suits for aircraft pilots, say, who may have to ditch in freezing waters? One view would be that the experiments were intrinsically ethically unacceptable and therefore their results should not be used for any purposes, no matter how important. The alternative opinion holds that, although the experiments themselves were completely unacceptable, their consequences may yet be beneficial. In other words, we should evaluate the ethical dimension here on the basis of the use to which these experiments can be put. And if they can be used to help save lives, then the appalling suffering of their subjects will not be in vain.

Or consider the current debate over animal research. One side insists that for many experiments the use of animals is necessary and these experiments and tests will have beneficial consequences for humanity. The other argues that they are unnecessary, even misleading, given the different physiologies of the animals involved and humans, and that the subsequent benefits do not outweigh the ethically abhorrent nature of the experiments themselves. It may be that legislation is passed which restricts certain kinds of experiments or outlaws them altogether, in which case certain scientific questions will go unanswered and certain developments will not be explored or undertaken. I'm not going to take a stance on these ethical issues here; the question is: does this undermine the objectivity of science? Again, the answer is surely not; ethical standards may constrain scientific practice in certain ways, just as funding or the lack thereof does, but within such constraints the nature of experimental results themselves and the content of theories remain unaffected.

3 Social factors may determine the *content* of scientific beliefs

Let's now turn our attention to the claim that social factors *cause* or bring about the kinds of theories scientists come up with and the ones that they believe in. Now we need to exercise just a little care before we plunge in. First of all, the idea that the discovery of scientific hypotheses and theories is driven by social factors may not be so problematic, particularly if you accept the separation between discovery and justification we discussed in Chapter 2. There, we recall, it was argued that theories may be discovered through all sorts of means but that what is important is how they are justified, or supported by the evidence. Even if the actual content of the theory is clearly determined by socio-economic or political factors – imagine that Darwin didn't travel on the Beagle, didn't study animal breeding and so on, but just reflected on Victorian society and came up with the idea of natural selection and survival of the fittest that way – it shouldn't matter in the long run as long as the theory is thrown to the wolves of experience and rejected or accepted on that basis. However, if that acceptance or rejection is biased by social factors – if, for example, what counts as evidence is so determined, or the impact of that evidence – then we might well conclude that the objectivity of science has been eroded, perhaps undermined altogether.

In such cases, we might well conclude that scientists have come to hold *irrational* beliefs. How, then, do we distinguish between rational (objective) and irrational (non-objective) beliefs?

The traditional answer distinguishes between rational and irrational beliefs precisely in terms of the influence of social factors: rational beliefs are held because they are true, justified by the evidence, etc. and hence are *objective*; irrational beliefs are held because of the influence of certain social factors. A well-known example of the latter from the history of biology would be the development of Trofim Lysenko's ideas in the old Soviet Union.

Lysenko was an agronomist from the Ukraine who was typically portrayed in the Soviet press as a kind of 'peasant scientist', more interested in practicalities than biological theory. He came to prominence through a technique he called 'vernalization', which allowed winter crops to be obtained from summer planting by soaking and chilling the germinated seeds. This offered hopes of a dramatic increase in agricultural productivity and provided the basis for Lysenko's theory that environmental interaction was more important for the development of an organism than genetic constitution. With geneticists under attack during the 1930s for their 'reactionary separation of theory and practice', Lysenko positioned himself as someone who had achieved practical successes, unlike the geneticists with their 'useless scholasticism'. Together with a member of the Communist Party, Isaak Izrailevich Prezent, Lysenko denounced genetics as,

> ... reactionary, bourgeois, idealist and formalist. It was held to be contrary to the Marxist philosophy of dialectical materialism. Its stress on the relative stability of the gene was supposedly a denial of dialectical development as well as an assault on materialism. Its emphasis on internality was thought to be a rejection of the interconnectedness of every aspect of nature. Its notion of the randomness and indirectness of mutation was held to undercut both the determinism of natural processes and man's ability to shape nature in a purposeful way.
>
> (H. Sheehan, *Marxism and the Philosophy of Science: A Critical History*, New Jersey: Humanities Press International, 1985, 1993).

In its place, Lysenko developed,

> ... a new theory of heredity that rejected the existence of genes and held that the basis of heredity did not lie in some special self-reproducing substance. On the contrary, the cell itself ... developed into an organism, and there was no part of it not subject to evolutionary development. Heredity was based on the interaction between the organism and its environment, through the internalisation of external conditions.
>
> (ibid.)

Hence according to Lysenkoism, there is no distinction between what biologists call the genotype, or the nexus of genes inherited by an individual, and the phenotype, that is, the characteristics of the individual that result from the interaction between heredity and the environment.

With genetics research slandered as being in the service of racism and caricatured as the 'handmaiden' of Nazi propaganda, and with leading geneticists arrested, imprisoned and even executed, Lysenko's theory came to be officially endorsed, with Lysenko himself quoting Engels (co-author with Marx of the *Communist Manifesto*) in support of it. The effects on Soviet genetics research and on biology in general were devastating, and it wasn't until the politically more tolerant mid-1960s that Lysenko was denounced, his theory rejected and his practical success revealed to have been ill-founded and exaggerated. In 1964, the physicist Andrei Sakharov stood up in the General Assembly of the Soviet Academy of Sciences and declared that Lysenko was

> ... responsible for the shameful backwardness of Soviet biology and of genetics in particular, for the dissemination of pseudo-scientific views, for adventurism, for the degra-dation of learning, and for the defamation, firing, arrest, even death, of many genuine scientists.
>
> Quoted here: http://www.learntoquestion.com/seevak/ groups/2003/sites/sakharov/AS/biography/dissent.html

Although an understandable desire to achieve practical successes played its part in this story (understandable since Soviet agriculture had suffered terribly from the forced collectivization of the 1920s), Lysenko's views were accepted and widely adopted on the basis of *political* considerations and hence this acceptance can be taken as unjustified and, ultimately, irrational.

An alternative answer to the above question of how we distinguish between rational and irrational beliefs throws the distinction itself into doubt and suggests that we should treat all beliefs as on a par, in the sense that so-called 'rational' and 'irrational' beliefs should be subject to the same kind of explanation, where, it turns out, that explanation will be in terms of social factors. So, rather than saying that certain beliefs are or should be held because they are true, or justified by the evidence, and others should not, this approach advocates equality of treatment – look at the social factors behind the acceptance of all beliefs, without exception. That may sound quite reasonable, but an advocate of answer 1 may well protest that it remains to be shown that the acceptance of scientific theories and hypotheses is driven by these social factors. What is needed, and what the defenders of the second answer above have provided in some cases, is a detailed reconstruction of particular cases of the acceptance of theories, explicitly indicating the factors involved and their impact. These studies have of course been disputed, but let's continue to explore this kind of approach.

If the content of theories is determined in this way, in the sense that they are not just discovered due to prevailing social conditions, but accepted for similar reasons, then our picture of science as objective, value-neutral, somehow standing above the socio-economic and political context has to be given up. Scientific theories and scientific 'facts' must now be seen as 'socially constructed'.

The social construction of scientific 'facts'

The view that theories are accepted, ultimately, for social reasons and that scientific facts are socially constructed has become generally known as 'social constructivism' and one of its most influential schools of thought is widely known as the 'Strong Programme'. The core idea of this position is that there is no reason why the *content* of *all* scientific beliefs cannot be explained in terms of social factors. It is founded on the following version of the above idea that we shouldn't introduce distinctions between rational beliefs, which are good, and irrational ones, which are in some sense bad:

The equivalence postulate: All beliefs are on a par with one another with respect to the causes of their credibility.

Here's how the two most famous advocates of the Strong Programme put it:

> The position we shall defend is that the incidence of all beliefs without exception calls for empirical investigation and must be accounted for by finding the specific, local causes of this credibility. This means that regardless of whether the sociologist evaluates a belief as true or rational, or as false and irrational, he must search for the causes of its credibility.
>
> (B. Barnes and D. Bloor, 'Relativism, Rationality and the Sociology of Knowledge', in M. Hollis and S. Lukes (eds), *Rationality and Relativism,* Cambridge, MA: MIT Press, 1982, p. 23)

By specific, local causes of credibility, here, Barnes and Bloor mean social factors, so the idea is to look for such factors behind the acceptance of all beliefs, without splitting them up into the rational and the irrational.

Now this raises a further interesting question: How is credibility established? In most cases, we don't get to work on that many scientific theories. Even if your particular theory, discovered, justified and accepted in the ways we have considered here, wins you a Nobel Prize, it is unlikely that you will have personally considered, evaluated and judged the evidence yourself. Typically we – whether scientists or lay-people – rely on the judgements of others, particularly experts in their fields. An important component of this reliance is obviously *trust*. So, this raises the further interesting question: Who or what can you trust?

One answer, which might be viewed as the traditional one, previously discussed in Chapter 6, is that you can trust the evidence of your own senses. It is this that supposedly grounds the objectivity of science. A more modern alternative is that you can trust the *experts*, but who are they? When you think of an expert you might immediately think of the TV image of the doctor or lab tech in the white coat but why should you trust someone in a white coat?! Well, it's supposed to reflect a particular social status, achieved after a certain level of training, and the person wearing it is supposed to emanate a certain level of trust. That's all fine and good when it comes to the iconography of TV ads but where does this leave objectivity? Again, the traditional view is that it leaves it exactly where it should be, since the expert is *transparently* objective. What this means is that the expert stands in a kind of chain, leading from someone making the observations to you, and all he or she does is to transmit the facts, as it were, along the chain, without adding to them or taking anything away, without distorting them or modifying them in any way. The objectivity we achieve from observation is passed on through the expert and that is why we can trust them, on this view: the expert is a transparent transmitter of the facts.

This is fine as long as we can be assured that the expert remains transparent and free from bias. But how plausible is this? The sociologist will insist that it is not very plausible at all, since he or she is immersed in a particular social context and hence will be subject to all the contingent social, political and cultural factors associated with that context. By shifting objectivity from the facts to the expert, immersed in a particular social context, it has become socially determined and hence, according to the traditional view, is not bias-free, social factor-free objectivity at all!

Now you might ask, can't we admit that trust *is* crucial to establishing scientists' knowledge claims but insist that facts still play a role? In this way, 'objectivity' might not be completely socially determined. Some sociologists reject even this insistence and claim that social factors determine the *content* of scientific beliefs. On this view, scientific 'facts' are nothing more than social artefacts or constructs. Researching the early history of modern science, Steven Shapin and Simon Shaffer focus on the work of Robert Boyle, today perhaps best-known for 'Boyle's Law' of gases (which basically states that the pressure exerted by an ideal gas is inversely proportional to the volume it occupies, if the temperature remains constant). In particular they examine his experimental work and they argue that this should be understood as an attempt to establish secure knowledge and scientific order in the context of the changing political order following the English Civil War. And radically, perhaps, they point to Boyle as the architect of the view that the scientist, who was of course a gentleman, should be seen as a modest witness whose apparently 'objective' language helped establish 'matters of fact' in the context of a community of like-minded individuals. They write,

> The objectivity of the experimental matter of fact was an artifact of certain forms of discourse and certain modes of social solidarity.
> (S. Shapin and S. Shaffer, *Leviathan and the Air-pump: Hobbes, Boyle, and the Experimental Life*, Princeton: Princeton University Press, 1985)

What could this mean? How are we to understand the claim that so-called scientific facts are 'socially constructed'? The sociologist's answer is that scientific knowledge is constructed through social interaction, that is, through a form of negotiation, between experts in laboratories. External reality is hence not seen as the cause of scientific knowledge; rather scientists establish 'reality' through the claims they make as purveyors of truth. This is quite a radical position to adopt and we can immediately appreciate that the social construction of facts leads to a form of *relativism*, since if the facts depend on the social context, then a different social context (at a different time, or in a different place) will lead to a different set of facts and different scientific knowledge.

Social constructivism and relativism

Let's look at this consequence a little more closely. One formulation of relativism takes it to hold that there is no privileged standard for the justification of beliefs. In other words, you cannot say that certain beliefs are justified and hence rational to hold because they are supported by the facts – what counts as a fact depends on the social context. Here's what Barnes and Bloor say:

> For the relativist there is no sense attached to the idea that some standards or beliefs are *really* rational as distinct from merely locally accepted as such.
> (B. Barnes and D. Bloor, 'Relativism, Rationality and the Sociology of Knowledge', in M. Hollis and S. Lukes (eds), *Rationality and Relativism*, Cambridge, MA: MIT Press, 1982, p. 27)

How do we arrive at such a position? Here are the three (easy) steps to relativism:

1   Different social groups hold different beliefs on a given issue;

2   What you believe is relative to the locally accepted standards of justification, i.e. the standards accepted by a specific social group (e.g. scientists, theologians, shamans);

3   Since there is no socially independent standard of justification, all beliefs are on a par.

According to the relativist, standards for the acceptability or justi-fication of scientific beliefs are socially determined by values that are external to science. There is no privileged 'global' justification, such as being in correspondence with the 'facts'. What counts as a scientific 'fact' is socially determined and so is the truth. Hence science is no 'better' than any other form of belief; all beliefs are equal because there is no valid distinction between what is 'really' objective 'knowledge' and what is locally accepted as such.

You might find such a view absurd and feel that if this relativism is a consequence of the sociological view of objectivity as deter-mined by social context, then the sociological approach must be rejected. However, Barnes and Bloor embrace their inner – and outer – relativist, declaiming,

> In the academic world relativism is everywhere abominated. Critics feel free to describe it by words such as 'pernicious' or portray it as a 'threatening tide'. On the political Right relativism is held to destroy the defences against Marxism and Totalitarianism. If knowledge is said to be relative to persons and places, culture or history, then is it not but a small step to concepts like 'Jewish physics'? On the Left, relativism is held to sap commitment, and the strength needed to overthrow the defences of the estab-lished order. How can the distorted vision of bourgeois science be denounced without a standpoint which is itself special and secure?
>
> The majority of critics of relativism subscribe to some version of *rationalism* and portray relativism as a threat to rational, scien-tific standards. It is, however, a convention of academic discourse

that might is not right. Numbers may favour the opposite position, but we shall show that the balance of argument favours a relativist theory of knowledge. Far from being a threat to the scientific understanding of forms of knowledge, relativism is required by it. Our claim is that relativism is essential to all those disciplines such as anthropology, sociology, the history of institutions and ideas, and even cognitive psychology, which account for the diversity of systems of knowledge, their distribution and the manner of their change. It is those who oppose relativism and who grant certain forms of knowledge a privileged status, who pose the real threat to a scientific understanding of knowledge and cognition.

(B. Barnes and D. Bloor, 'Relativism, Rationality and the Sociology of Knowledge', in M. Hollis and S. Lukes (eds), *Rationality and Relativism,* Cambridge, MA: MIT Press, 1982, pp. 21–2)

That is, they insist that relativism is *essential* for understanding how science works.

Now this is a provocative view, but it faces certain problems:

Problem 1 If all views are relative to social context, what about relativism itself?! If the defenders of relativism and the social construction of scientific facts insist that their view is objectively correct, then their relativism is selective. However, there is a straightforward response to this: the belief that scientific 'facts' are determined by social factors is *itself* determined by social factors. This is known as 'reflexivity'; the relativist is reflexive in holding that relativism is itself relative. Of course what that means is that you could always respond that social factors lead you to maintain that scientific facts are *not* determined by social factors, but in accepting that your claim that science is objective cannot itself be objectively defended, you've given the game away!

Problem 2 Relativism blocks change, both political as well as scientific. This is potentially a more serious problem. Consider: If what counts as a fact is determined by and relative to political context then communist-biased facts, say, are just as 'objective' on this view as non-communist or 'capitalist-biased' ones and only a change in political context will lead to any change in the relevant science. If what counts as acceptable science is determined by the local community, then if there's no change in the social community there will be no change in science. If we have no objectively justified, rational grounds for accepting one theory rather than another, in what sense can there be scientific progress? Well, we shall return to this question in our final chapter, but let me say here that the standard position is that progress is towards the truth and is driven by objective, rational factors such as those having to do with the evidence. If we accept some form of relativism then this form of progress goes out the window; and all we have left is change through change of social context. If we think there *is* scientific progress in the standard sense, then we'd be inclined to reject the relativist position.

Problem 3 Relativism blocks communication and understanding, whether between different social communities across the world today or between different scientific eras across time. The fact that we can understand the beliefs of cultures which are very different from ours and of scientists from the eighteenth, seventeenth or pre-AD centuries is surely indicative that not *all* beliefs are relative; different communities (whether scientific, cultural, or whatever) *share* some common beliefs. It is in this vein that the rationalist philosopher Lukes insisted that,

> ... the existence of a common reality is a necessary precondition of our understanding [another society's] language.
>
> (S. Lukes, Some Problems About Rationality', p. 209, in
> B. Wilson (ed.), *Rationality*, New York: Harper & Row, 1970,
> pp. 193–213)

What he means by this is not that we must agree on the reality of quantum fields, or curved space-time, for example. What he means is that this other society must possess our distinction between *truth* and *falsity*, because if it did not '... we would be unable even to agree about what counts as the successful identification of public (spatio-temporally located) objects' (ibid., p. 209). It is on the basis of such an agreement that a kind of 'bridgehead' between the two cultures or two scientific eras can be constructed. So the idea is that we can begin to piece together an understanding of what Newton or Darwin or Freud believed because they shared our distinction between 'true' and 'false' at the basic level of the kinds of objects we can see with our own eyes all around us. And likewise we can begin to understand the beliefs of different scientific communities today, even if some of those beliefs are driven by, say, political considerations. So, consider the Lysenko case again: even though Lysenko's theories were developed and widely accepted on the basis of political factors, opponents were still able to understand them, debate them and, to their cost in many cases, reject them.

Now the relativist might respond that even when it comes to 'public' objects, people in other cultures, other societies may have very different beliefs about them. Take a large hill, for example, standing prominently out in the countryside: we might see it as a particular geological formation, but the local culture might view it as a source of magic, or the home of a sleeping king and his retinue who will awake to defend the country in its hour of need. So, there is a form of relativism even here. However, the rationalist is not insisting that members of the other society must identify a hill as a 'hill', in the sense that we do (as a geological formation, say), but that they must be capable of distinguishing it from a tree, for example, or a pool of water. Although the members of the other society may attribute certain properties to hills that we do not – such as possessing magical powers, say – they must attribute enough of the properties that we do in order to distinguish a hill from a small pool of water, say. Such a property might be that of relative impenetrability, so that our friends from the other society will agree in assigning the value 'True' to the statement 'You can't walk through a hill.' As far as the anti-relativist is concerned, that is all we need to start building our bridgehead with this strange other culture.

That this kind of bridgehead can be built is supported in two ways. First of all, there is the evidence from anthropologists themselves, who go to strange, far-away places, study cultures very different from ours and observe people studiously not walking through hills. In other words, despite what the relativist says, there are apparently no cases of anthropologists, or historians of science for that matter, returning from their studies empty handed and saying, 'Nope, I just couldn't understand that social or scientific community at all.' However, this kind of practice-based reason is not entirely straightforward, since the relativist may counter that neither the anthropologists nor the historians approach the other culture or scientific era with a kind of blank slate; rather, they have their own philosophical predilections which they may bring into play, particularly if they have received training themselves in the methods of science and hence our framework of rationality. So, it is not surprising that they return with reports of contact of a form that supports such a framework. In other words, the data the anthropologists or historians bring back with them are already laden with our society's (broad) theory of rationality.

We recall our earlier discussion of the theory ladenness of observation in Chapter 7. It can be tackled in two ways: first, the theory with which the observations are laden is not the theory being tested; second, observations in science are typically 'robust' in the sense that the data remain (broadly) the same across a range of instrumentation with correspondingly different background theories. The first response is not available to the rationalist, since the concern is precisely that data are laden with exactly the theoretical framework that is being tested. The second response is potentially very interesting but what would be required would be for anthropologists or historians of science from radically different cultural backgrounds to ours to make the necessary observations, and it is hard to see how that could be achieved. Here the rationalist may well throw up her arms and insist that now questions are being begged against her, since she insists that there could be no such radically different cultural backgrounds!

There is also a second reason that is particularly interesting with regard to our discussion of rationality and objectivity in science. Lukes expresses it as follows:

> ... any culture, scientific or not, which engages in successful prediction (and it is difficult to see how any society could survive which did not) must presuppose a given reality [and] ... it is, so to speak, no accident that the predictions of both primitive and modern common-sense and of science come off. Prediction would be absurd unless there were events to predict.
>
> (ibid, p. 209)

Now this appears to be nothing more than a form of the infamous 'No Miracles' argument, which underpins scientific realism and which we discussed in Chapter 8. The gist of the argument, we recall, is that the success of science – where this is understood in terms of making predictions – would be a miracle, unless the claims it makes about reality were (broadly) true and the objects it posits also exist. However, as we noted previously, this is highly contentious, and again, we recall, it is typically presented as a form of inference that the anti-realist rejects as deeply problematic and, indeed, question begging. Of course, as applied to their project of defending rationalism against the relativists, rationalists might reply that such objections are all very well when it comes to the unobservable objects of science, but that what they are concerned with, of course, is everyday reality and its 'public, spatio-temporally located' objects, like rocks and hills. So, consider the following example: the best explanation for the noise behind the skirting board, the nibbled biscuits etc. is that there is a mouse in the house; therefore there is a mouse in the house. If the relativist is willing to accept this for 'everyday' objects, such as mice, which sit on the bridgehead, then she should be willing to accept the more general form as suggested by Lukes above and her opposition to universal criteria of rationality and objectivity would be undermined.

However, as often is the case in philosophical debates, things are not quite so simple, unfortunately. In the context of the realist–anti-realist debate, the constructive empiricist – whose position we discussed in Chapter 9 – will reject the move from accepting these kinds of inference for everyday objects, such a mice, to accepting them in general. And although the constructive empiricist is no relativist, they will typically insist that we have to be careful about the form of rationality we put up in opposition to the relativist's view. In particular, they may reject what has been called a 'Prussian' rule-based approach to rationality in favour of an 'English' permissive stance. According to the former, you are rational *only if* you follow certain rules, whereas on the latter view, you are rational *unless* you violate certain constraints, such as being consistent (so, let me ask: was Bohr rational in proposing his famously inconsistent theory of the atom?!).

The point is that it is not clear that the rationalist can justify the above reason for rejecting relativism, as stated by Lukes, on the grounds that we typically apply it locally in the form of inference to the best explanation, even when it is only 'public' objects that we are concerned about. His relativist opposition might insist that since we are not *compelled* to accept it at the local level, we are likewise not compelled to adopt it as applied globally.

Even worse, the relativist may feel that the whole argument, and in particular the claim that 'prediction would be absurd unless there were events to predict', begs the very question at issue. The claim that successful prediction would be a miracle or absurd, unless the primitive, common-sense or scientific claims from which such predictions are drawn actually 'matched' reality in some way, presupposes a view of the relative plausibility of miracles which may be particular to our culture. Other cultures may have beliefs according to which the occurrence of miracles is not so implausible and hence in that context successful predictions, even at the low level of planting crops and avoiding walking into hills, might indeed be regarded as accidental or miraculous.

At this stage of the debate, the rationalist might well protest that the relativist is *not herself* a member of a society in which miracles are taken to occur on a regular basis, but a member of *ours* in which they are not. If the relativist refuses to accept the very framework of debate and argument of *our* culture, then what is the point of any further discussion? Of course, the relativist may insist that she does indeed accept the rules of academic debate but only as contextually determined. In that case, however, the rationalist may feel that her argument goes through, since the point is not whether members of another, so-called 'primitive', society accept the argument, but whether *we* do. And if we do, even if only on a contextual basis, then relativism is undermined – only in *this* context, granted, but then, that's the only context that is relevant for these purposes!

At this point we must stop. We have moved far from our original concern with objectivity, rationality and the independence of science from social and political factors. Like many philosophical debates the matter has not been decisively settled but I hope you now have some idea of the issues at stake. We can highlight some of these issues even further by considering a concrete example of the influence of certain factors such as gender bias, for example, and that is the topic of the next chapter.

11

Gender bias

Introduction

In the previous chapter we considered whether social factors in general might undermine the objectivity of science and we looked at one view which maintains that they do. Here we will look at one particular factor, or set of factors, and consider the extent to which they threaten that objectivity. The factor concerned has to do with the tricky issue of gender, so our fundamental question will be: Does gender bias undermine the objectivity of science?

Science as an androcentric activity

Let us consider the ways in which gender bias might impact on science.

1 **Gender bias may determine the *proportion* of men and women in science**
   This seems entirely plausible. Recent surveys have concluded that although, on average, women account for 53 per cent of the 'professional' labour force, across the European Union, compared to just 45 per cent of the total labour force, only 32 per cent of science and engineering posts were held by women (see http://ec.europa.eu/research/science-society/document_library/pdf_06/she-figures-2012_en.pdf). Previously, a major cause of gender imbalance in

science was lack of education: women were either actively discouraged from pursuing science degrees or at the very least dismissed as oddities or jokes. Since the mid-60s, however, the number of women receiving bachelor's degrees in science and engineering has increased year on year, and is now roughly half of the total. Nevertheless, it is clear that women face certain gender-related barriers to entry into a scientific career and, further, to success in such a career: it is now well known that the proportion of women with higher grades in academia, for example, decreases quite dramatically as we move up the ladder towards the professorships.

It's not clear what the reasons are. Typically childcare issues are raised: women are not hired as scientists because they are seen as a risk insofar as they may leave if they become pregnant; women have difficulty returning into the profession after giving birth, because of arranging for suitable childcare; or women have difficulty establishing themselves as part of the scientific team because they have family obligations that prevent them staying until all hours in the lab. An EU report put it as follows:

> Reasons for this imbalance are multiple. Certain fields are considered to be men's property and, therefore, gender bias affects judgements on scientific excellence. Industries and academia are also reluctant to hire women because they are not seen as flexible enough. Employers also fear that women may choose to give up their careers and start a family instead.
> (http://www.euractiv.com/en/science/women-science/article-143887)

As deplorable as this situation is (and it has certain practical consequences – the lack of women scientists is cited as one of the reasons the EU is having difficulty in meeting its goals of becoming 'the world's most competitive knowledge-based economy'), our answer to the question, 'Does this undermine the objectivity of science?', is surely 'No'. It may be unproductive and inefficient but the existence of gender bias of this sort provides no grounds for thinking that the content of theories or the conduct of scientific experiments is in any way affected.

2   **Gender bias may determine *what* science investigates**
This is not just to suggest that certain scientific topics are deemed to be too 'girly' (although that might well be the case) but that the allocation of resources, both human and financial, may be determined or, at the very least, influenced by considerations of gender. So, for example, take the case of contraception: generally it has been the case for many years now that the bulk of research into contraceptive techniques, devices, drugs etc. has focused on the female side of things. The female birth control pill was introduced in the early 60s, but despite being repeatedly reported as 'just around the corner', an 'easy-to-use' contraceptive pill for men is still not available at the time of writing (in 2015; see: http://www.theguardian.com/society/2014/feb/01/who-wants-male-contraceptive-pill-chauvinism). And more generally, of the thirteen methods of contraception currently available, only three are aimed at men; the rest all leave it to the woman to take responsibility.

Compare the development of both the female and male pill with the development of 'Viagra' and other medications which address the problem of erectile dysfunction. Pharmaceutical companies were reluctant to develop and market the female contraceptive pill, primarily because of restrictive birth control laws in place at the time, whereas 'Viagra' was developed and actively marketed comparatively quickly. It has also been claimed that the way side-effects of these two drugs are publicized and dealt with is indicative of some form of gender bias. So, widely publicized reports of deaths led to the calls for the pill to be withdrawn in the USA, whereas mortalities associated with Viagra have not made the headlines. On the other hand, it has also been suggested that health problems such as strokes etc., associated with the early high doses of the female pill, were not addressed by the medical profession and it took years of campaigning and pressure before a low-dose version of the pill was developed. So, it's not always clear where the gender bias bites.

Nevertheless, what is perceived as an inherent gender bias in family planning services is often the target of criticism by women's health groups who point out that women typically bear the burden of contraceptive use, largely because of this bias. However, if a wide ranging comparison is undertaken, covering such issues as health risks, side-effects, effectiveness and, of course, convenience, then male-oriented contraception, such as condoms and vasectomy, scores highly and would obviously take some of that burden off women.

Again, we must ask the question: Does this undermine the objectivity of science? No. As uncomfortable as we may be both with the topic and the findings, this appears to be another case in which social factors constrain the science that is done, but that science is still done in an objective manner.

3   Gender bias may determine *how* science investigates
Let's move on and start to consider ways in which objectivity might be threatened. Consider the claim, for example, that 'virtually all of the animal-learning research on rats has been performed on male rats'. First of all, this may be seen as reflecting an attitude of the male being the norm, extended to other species, with the female seen as a deviation from this norm. But second, concerns may arise over extrapolating from such experiments to humans, and in particular women. You might have such concerns anyway, and as we touched on in the previous chapter, animal experimentation has been criticized on the ground that the difference in physiologies, for example, blocks any such extrapolation. But further worries might arise with regard to the legitimacy of extrapolating from experiments performed on the males of one species, to both male and female humans.

Consider the example of research into strokes (where injury may result from interruption of the blood supply to certain parts of the brain): it is well known that there are numerous gender-related differences with regard to the mechanism of strokes, and their impact. These differences have an important bearing on the patient's responsiveness to the various kinds of treatment, and hence may seriously affect survivability, recovery and subsequent quality of life. Yet most of the animals used in stroke research are male and the results obtained are often extrapolated to women, despite the above differences. Of course, the researchers involved may insist that there is no explicit gender bias; they may argue that male rats are easier to handle, for example, since they don't get pregnant and aren't subject to the same hormonal changes. But that just underscores the point: unrepresentative male models are used to conduct scientific research which is then expected to apply to women. What is required are female specific models.

Or consider research into cardiovascular disease in general (this includes coronary heart disease, stroke and other cardiovascular diseases). For many years this was regarded as primarily a male-oriented disease, and it is only recently that awareness has grown that it is a major health issue for women as well. In 2009, women accounted for 51 per cent of deaths from cardiovascular disease in the USA and in the UK in 2012 it accounted for 28 per cent of all deaths among women, more than for cancer. For young women in the UK, cardiovascular disease is a bigger killer than breast cancer (see: http://medicalxpress.com/news/2015-06-cancer-cardiovascular-disease-uk-killerbut.html).

Despite this, research into cardiovascular disease is heavily skewed towards male subjects and male concerns. Thus, a study in the Journal of the American Medical Association from 1992 found that women were excluded from 80 per cent of the trials for acute myocardial infarction, more commonly known as a heart attack. The report concluded that the results of the trials could not be generalized to the female patient population. Furthermore, in considering the treatment of the disease, the doses of drugs given to both men and women with heart disease are often based on studies of, typically, middle-aged men even though it is known that women generally suffer from cardiovascular disease at an average older age than men, that women generally have a smaller overall body mass than men and that male and female hormones are, of course, different, all of which may affect drug concentrations, overall effectiveness and side-effects. And finally, research over the past twenty years or so has demonstrated the beneficial effects of aspirin as a preventative medication, yet women were excluded from an early major study into this effect. More recent large-scale research in the form of the Women's Health Study in the USA has shown that regular low-dose use of aspirin reduced the risk of stroke by 17 per cent, but did not decrease heart attacks or cardiovascular deaths among all women.

Now, here we might well be concerned about the objectivity of science. If the theories about, say, the effectiveness of aspirin in preventing cardiovascular disease are taken to apply to both men and women, but the evidence justifying them is obtained from studies in which women were excluded, then there would appear to be a clear evidential bias here. Of course, the bias can and, it appears, was corrected, with women subsequently included in such studies, and one might argue that objectivity was restored. Let's now look at cases where the undermining of objectivity is apparently much more severe and the possibility of correction much more difficult.

4   **Gender bias may determine the *content* of scientific beliefs**

This seems, on the face of it, a radical claim. It is not suggesting merely that the evidence may be skewed, as in the above cases, but that the actual *content* of scientific theories – what they supposedly say about the world – may be gender biased.

Let's consider an example which may make such a claim plausible. This is taken from the field of primatology, the study of non-human primates. Putting things a bit crudely,

the story goes like this: in the 1930s, 40s and 50s male
primatologists went out into the jungle, observed monkeys
and apes and came back with theories about ape and
monkey behaviour which emphasized the dominance of
the males and the subservience of the females and which
both fitted with and helped further support certain social
beliefs about the role of women in 'Western' society. Then,
following the increase in women university graduates in
the 1960s and 1970s, female primatologists headed off into
those same jungles, also observed the apes and monkeys
and came back with very different observations, supporting
very different theories. So, whereas the male primatologists
had portrayed groups of primates as consisting of basically
a dominant male and his 'harem' of submissive females and
had paid particular attention to male aggressive behaviour,
women scientists re-described this in terms of procreational
advantages for the female primates, according to which
males are simply a resource which the female may use to
further the survival of herself and her offspring. From this
perspective, the group needs only the one male because
his sole role is to impregnate them. Even the language is
altered, with 'harem', and its historical associations, replaced
by the more neutral phrase 'single-male troop of animals'
(the classic account of this shift is given in *Primate Visions:
Gender, Race and Nature in the World of Modern Science*, by
Donna Haraway, New York and London: Routledge, 1989; and
more recent discussions can be found in Shirley C. Strum
and Linda Marie Fedigan (eds), *Primate Encounters: Models
of Science, Gender, and Society*, Chicago: University of
Chicago, 2000).

Female scientists not only questioned the theories
developed by male primatologists (and the way they were
used to bolster the current stereotypes of male–female
relationships and behaviour), but they also criticized the
observational techniques used, such as the sampling
methods, which were very much male centred. They also
subjected classic fieldwork to critical analysis and questioned
the extrapolations that had been drawn, not only to human

behaviour, but even that of other primates. In the late 1920s and early 30s, for example, Solly Zuckerman (who subsequently became chief scientific advisor to the British Government) undertook a widely publicized study of captive hamadryas baboons, in which he observed extensive male violence, with male baboons attacking and killing each other in large numbers. This reinforced the picture of primate groups as involved in life or death struggles for male dominance, with, again, the usual extrapolations to human society.

However, the study was subsequently criticized for its bias and the unrealistic conditions under which the baboons were kept. It was pointed out that they were massively overcrowded with a male to female ratio that was very different from what it would be in the wild. Furthermore, hamadryas baboon females turn out to be among the most submissive and most gender-unequal of all primates. Together with the influx of women scientists into the field, these criticisms helped to shift attention to other primates that showed very different behaviour patterns and 'troop' structures, in many of which the females held social control.

Here we see, then, how gender bias can creep into the observations made, the conclusions drawn and the questions asked, and thus into the content of the theories produced. Does this undermine the objectivity of science? It would appear so, and here's another example.

Case study: Gender bias in paleoanthropology

Paleoanthropology is the study of fossilized human beings. The earliest hominids lived on African savannahs, at least 3.4 million years ago; they were about the size of modern-day chimps but had slightly larger brains and walked upright. It has been argued that the content of theories of early hominid evolution has been determined by certain gender-based assumptions. Let's examine this argument more closely.

First of all, let's consider the phenomena that the theory of hominid evolution tries to explain. There are basically three developments that are regarded as crucial for the evolutionary development of our early ancestors. These are:

1 Increased cranial capacity (larger brains);

2 Tool use (the development and use of stone tools);

3 Bipedalism (the shift from walking on all fours to walking on two legs).

Two theories have been put forward, which have been labelled 'Man the Hunter' and 'Woman the Gatherer', and they offer contrasting accounts of the above phenomena.

Theory 1: Man the Hunter

The core idea here is that hunting was the driving force behind human evolution:

> The biology, psychology, and customs that separate us from the apes – all these we owe to hunters of time past ... for those who would understand the origin of human behaviour there is no choice but to understand 'Man the Hunter'.
> (S. L. Washburn and C. S. Lancaster, 'The Evolution of Hunting', in *Man the Hunter*, R. B. Lee and I. DeVore (eds), Chicago: Aldine Press, 1968)

Now obviously we cannot go back in time and observe early hominid societies to determine if it really was hunting that led to the above developments. So, a different methodology has to be chosen and what scientists do is try to relate contemporary hunter-gatherer societies and wild primates to the fossil evidence.

Following this approach, the following explanations can be constructed:

1 The demands of hunting led to the development of communication and close interaction between members

of the hunting group and this favoured the development of larger brains;

2 The hunting and subsequent butchering of the kill led to the development and use of stone tools;

3 Hunting and tool use led to selective pressure to free the hands and this led to bipedalism.

The conclusion, then, is that the Man the Hunter theory nicely accounts for these crucial evolutionary milestones. Note the central assumption, that the changing behaviour of one sex (male) constitutes the centrally important adaptive strategies for early hominid evolution. Again we ask our question: Is this an example of gender bias? Let's consider the alternative.

Theory 2: Woman the Gatherer

The above picture of human evolution as due to the behaviour of one half of the species was strongly criticized by female paleoanthropologists:

> So, while the males were out hunting, developing all their skills, learning to cooperate, inventing language, inventing art, creating tools and weapons, the poor dependent females were sitting back at the home base having one child after another and waiting for the males to bring home the bacon. While this reconstruction is certainly ingenious, it gives one the decided impression that only half the species – the male half – did any evolving. In addition to containing a number of logical gaps, the argument becomes somewhat doubtful in the light of modern knowledge of genetics and primate behavior.
>
> (S. Slocum, 'Woman the Gatherer: Male Bias in Anthropology', in *Toward an Anthropology of Women*, by R. R. Reiter (ed.), New York: Monthly Review Press, 1975, p. 42)

The alternative Woman the Gatherer view adopts the core idea that *gathering* was the driving force behind human evolution. The methodology is broadly the same as above: that is, the division of labour between men hunting and women gathering is accepted, but the evolutionary weight is shifted from one to the other.

The explanations of the crucial developmental changes are now quite different:

1   The demands of gathering, of finding, identifying and picking the fruits, nuts etc., led to the development of co-operation and social organization and this in turn sparked the development of larger brains;

2   The gathering and cracking of nuts, seeds etc. led to the development and use of stone tools;

3   Gathering and tool use led to selective pressure to free the hands and hence to bipedalism.

The conclusion then is that the Woman the Gatherer theory also accounts for the various evolutionary milestones quite nicely. Note the common assumption in both accounts, namely that the contemporary hunter and gatherer societies, and monkeys and apes, are relevantly similar to early hominids. We have already seen that the latter part of this assumption may be questionable, as the case of the hamadryas baboons suggests they are not an appropriate model for early hominids. Likewise, different varieties of chimps show radically different social behaviour and it is not clear to what extent this behaviour can be extrapolated back in time and across the species.

So what we have here is another nice example of the underdetermination of theories by data:

> Any serious reconstruction of the past must 'fit' within a growing body of data on living apes and gathering-hunting peoples, the hominid fossil record, genetic relationships of living species, as well as concepts of evolutionary biology.
>
> (A. Zihlman, 'Women as Shapers of Human Adaptation', in F. Dahlberg (ed.), *Woman the Gatherer*, New Haven: Yale University Press, 1981)

The problem is, both theories appear to 'fit' the data!

What about further data that might break the underdetermination? There has been some interesting further research which has looked closely at the pattern of tool marks on animal bones and which suggests that early hominids were actually scavenging rather than hunting (see 'Man's early hunting role in doubt', *New Scientist*, January 2003). By comparison with modern day human scavengers (recall the methodology above), scientists have concluded that the meat supply would not have been enough to live on, and this has been taken to undermine the Man the Hunter theory and support the Woman the Gatherer. That's not to say that hunting didn't play *some* role in human evolution, however, and perhaps some combination of the two theories is the appropriate way forward.

Now, how should we respond to this situation?

Feminist responses

Some feminist commentators have argued that what the above demonstrates is that *all* the 'content' of theories is gendered and that there is no 'objective' way of selecting one theory over the other. Objectivity itself is nothing but a masculine ideal and should be rejected. Theory acceptance is relative to the social context and social factors must be acknowledged as driving theory choice. This is quite a radical view, and it raises a number of problems.

First of all, it obviously adopts a form of relativism, as we discussed in the previous chapter. But this is a dangerous move if one is a feminist hoping to *change* the way science is conducted, since it nullifies potential for any such change. After all, a male (non-feminist) scientist could simply insist that he chooses the Man the Hunter theory because it fits better with his social context and there is no further factor the feminist can appeal to in order to persuade him otherwise. If even the evidence is gender biased, then how can it be used to support the choice of one theory over the other. The problem, then, is that the relativism associated with this radical approach may undermine the broadly social and political aims of the feminist.

The second problem has to do with the plausibility of bias when applied to theories beyond anthropology and primatology, where some subjective element in observation may be unavoidable. Even if theories of early hominid evolution, for example, are subject to gender bias, this does not mean that *all* theories are, such as theories in chemistry, engineering, physics. When it comes to the latter, claims of gender bias seem a lot weaker. Here, there is no subjective door through which such bias can enter, either in observations or at the level of theory. Take General Relativity, for example: in what sense can this be said to be biased by gender? Of course, it was developed and elaborated by a man (and there have been claims that Einstein's earlier work on the Special Theory of Relativity owed more to his then wife's input than had previously been appreciated: claims which have since been rejected) but it doesn't seem that Einstein's maleness affected in any way either the content of the theory or its experimental confirmation. Accusations are sometimes made that the *interpretation* of quantum theory is somehow inherently gendered, or that the reductionist attitude often associated with it (in the sense that chemical bonding gets explained in terms of quantum physics and hence chemistry can be considered to be reduced to physics) is a product of a male-oriented society, but these are clearly pretty feeble!

A less radical approach is to admit that *some* of the content of scientific theories is gender biased and that an examination of science (as in the case study above) can reveal this bias by exposing the background assumptions of the relevant community. These assumptions are typically 'invisible' to that community and so we need alternative perspectives in order to expose and criticize them. Helen Longino is a well-known feminist philosopher of science who argues for something like this approach:

> The greater the number of different points of view included in a given community, the more likely it is that its scientific practice will be objective, that is, that it will result in descriptions and explanations of natural processes that are more reliable in the sense of less characterized by idiosyncratic subjective preferences of community members than would otherwise be the case.
>
> (H. E. Longino, *Science as Social Knowledge: Values and Objectivity in Scientific Inquiry*, Princeton: Princeton University Press, 1989, p. 80)

So, the idea is that if different communities are held accountable to each other, bias will effectively be weeded out. Of course, a problem here is that if one community refuses to even recognize the bias, then very little weeding will take place. What is required, insists the more radical feminist, is some form of 'oppositional consciousness', whereby feminist political activity is drawn upon to challenge gender bias within science.

Furthermore, the obvious question to ask is: What is the sense of 'objective' in the above quote? If by 'objective' we mean that scientific practice is independent of its social context then we are back to the traditional view. If, on the other hand, what is 'objective' is understood as relative to a given community then we return to a form of relativism. Perhaps, then, there is no middle ground and we are simply forced to choose one understanding over the other.

Conclusion

This seems as good a place as any to draw these discussions to a close, not just for this chapter, but for the book as a whole. I hope that the reader can see that the above considerations of gender bias are the concluding manifestations of a general theme running throughout this book, which has to do with the objectivity and rationality of science. We began with the process of discovery and I tried to lay out an alternative conception to the usual picture – often painted by scientists themselves – of scientific insights illuminated when the lightbulb goes off, a conception that emphasizes the rationale behind discoveries as represented in the heuristic moves that are made. This approach was pursued into the 'domain' of justification where I followed those philosophers of science who have suggested that a form of objectivity can be *achieved* (note the emphasis) even in the face of the apparent loss of security of the 'observational base' that underpins the testing process. We then considered the issues that cluster around the relationship between theories and the world, before tackling those views of science which hold that the relationship is actually between theories, or science in general, and their socio-politico-economic context.

I realize that I have not given definitive responses to the arguments of the relativists or the social contextualists, or even perhaps anywhere near adequate responses, but I hope that I have both outlined the central issues involved and given some indication of their complexity. Things aren't cut and dried here and nor should they be; science is a multi-faceted, complicated and, ultimately, intriguing phenomenon and if I've convinced you of that, and also that it is worth understanding in all its nuanced complexity, then my job here is done!

Exercise 5

Based on Chapters 9, 10 and 11, here are the questions to help you work through the topics:

Q1 What is meant by 'constructive empiricism'? Why is this described as a kind of 'anti-realism'?

Q2 What is 'entity realism'? What is someone who adopts such a view a realist about?

Q3 How does entity realism claim to overcome the Pessimistic Meta-Induction? Is it successful in doing this?

Q4 What features of science motivate entity realism? Can entity realism explain all aspects of 'miraculous' empirical success?

Q5 What is 'structural realism'? What is someone who adopts this position a realist about? How does it compare to 'entity realism'?

Q6 How does structural realism claim to overcome the Pessimistic Meta-Induction? Is it successful in doing this?

Q7 Which of these three views – constructive empiricism, entity realism or structural realism – do you favour? And why? If you prefer none of them, explain why and indicate what alternative position you would adopt.

Q8 What does it mean to claim that scientific facts are 'socially constructed'? How does social constructivism challenge the traditional conception of scientific objectivity?

Q9 What is 'relativism'? How does social constructivism arguably lead to relativism? Why is this problematic?

And here are some that should take you deeper into the issues:

A   What does the constructive empiricist mean by 'empirical adequacy'?

B   How does entity realism cope with the underdetermination problem?

C   What challenges does entity realism face?

D   What kind of continuity does structural realism find in the sequence of theories of light from classical to quantum physics?

E   What challenges does structural realism face?

F   How can gender bias affect the objectivity of science? Can it also affect the *content* of scientific beliefs?

12

Summary: Where we've been and where to go for more

When I teach my intro-level course I tend to present the topics in terms of questions and answers on the handouts. And at the end I offer a quick summary in the form of a Q & A revision guide. So I thought I would reproduce that here, followed by a bibliography of additional readings in the philosophy of science.

Discovery

Q: Is scientific discovery irrational or subjective?

A1: Yes, according to the 'Eureka' view (but no problem if we make the discovery-justification distinction).

A2: No, according to the inductive account (but there is more to discovery than observation).

A3: No, according to the 'heuristics' approach (there are rational moves behind discovery).

Justification

Q: How do scientific theories explain things?

A1: By deducing a statement about the phenomena from laws (plus background conditions).

A2: By giving the relevant cause of the phenomenon.

A3: By unifying different theories.

A4: By all of the above, depending on the context, and more!

Q: How are scientific theories objectively justified?

A1: Through observation (problem: there's more to seeing than meets the eyeball).

A2: Through a complex process of observations, interventions etc. (objectivity is an achievement).

Q: What is the nature of this relationship?

A1: Observations verify theories (problems: Which bits? How much?).

A2: Observations falsify theories (problems: Which bits? Theories are born falsified).

A3: It's a complex process of verification and falsification in which theories and observations are brought together in a variety of ways, including and often via *models*.

Realism

Q: What do theories tell us about the (objective) world?

A1: Theories tells us how the world *is* (standard realism):

- theories tell us about entities (entity realism),
- theories tell us about structures (structural realism).

A2: Theories tell us about how the world *could be* (constructive empiricism).

Independence

Q: Is science independent of its social context?

A1: Of course not! (trivial – no threat to objectivity):

- science is *dependent* on social context.

A2: Of course not! (non-trivial – undermines objectivity):

- science is *determined* by social context,

- this leads to relativism (of gender, class, race, culture ...),

- and a third answer to the questions under realism above: theories tell us about the social conditions under which they are constructed (social constructivism).

Further reading

The following are introductory and/or accessible texts that offer, in some cases, overlapping, in others, complementary perspectives to the one I've given here (some of these are adapted from the list given in S. French and J. Saatsi (2011/14), *The Bloomsbury Companion to the Philosophy of Science,* London: Bloomsbury (also included below), with thanks to my co-editor Juha Saatsi).

Bird, A., *Philosophy of Science (Fundamentals of Philosophy),* London: Routledge, 2002.

Bird begins his introductory book with the question 'what is science?', using as an example the debate between creationism and evolutionary theory. The rest of the book is then divided into two parts: 'Representation', covering laws of nature, explanation and realism; and 'Reason', which deals with induction, probability and scientific progress. Accessible and well written, the book also includes a glossary of terms.

Bortolotti, L., *An Introduction to the Philosophy of Science*, Cambridge: Polity Press, 2008.

This is another accessible and clearly written introduction to a wide range of issues, from how to demarcate science from pseudo-science, to the relationship between language and reality and the nature of scientific rationality. Unusually for an introductory textbook in the philosophy of science it includes a chapter on the ethics of science, as well as a glossary and a list of useful resources.

Carnap, R., *An Introduction to the Philosophy of Science*, New York: Dover Books, 1995/66.

A bit of a historical artefact and so also outdated (the clue is in the publication date!), it nevertheless offers a highly readable intro-ductory account of a range of fundamental issues, from explanation and prediction, through experiment and measurement to issues in the foundations of quantum physics and space-time theory, all from the perspective of one of the founders of logical positivism and one of the greatest philosophers of science of all time (and an outstanding member of the human race in general!).

Chalmers, A., *What is this Thing called Science?*, Brisbane: University of Queensland Press, 1999/76.

Although this is an accessible introduction that covers a range of core issues in the philosophy of science, it's now a little out of date on a number of issues. It begins with the common-sense view that science is based on 'the facts', through to a useful discussion of observation and the role of experiment before discussing induction, falsificationism, Kuhn's paradigm-centred philosophy of science, Lakato's methodology of scientific research programmes and Feyerabend's 'anything goes' approach.

Gillies, D., *Philosophy of Science in the Twentieth Century: Four Central Themes*, Oxford: Blackwell, 1993.

Gillies chooses as his four central themes: 'Inductivism and its

Critics', 'Conventionalism and the Duhem-Quine Thesis', 'The Nature of Observation' and 'The Demarcation Between Science and Metaphysics'. As should be obvious, there is a strong historical orientation and as a result it might be felt that these themes are not quite so central any more. Nevertheless, this offers a detailed and insightful analysis of the development of the philosophy of science in the twentieth century.

Godfrey-Smith, P., *Theory and Reality: An Introduction to the Philosophy of Science*, Chicago: University of Chicago Press, 2003.

Although like many introductory texts this book is set out chronologically, covering the last hundred years or so in the philosophy of science, Godfrey-Smith notes in the introduction that the chapters can also be read out of order, thematically. The usual suspects feature, from empiricism and falsificationism, to Kuhn and the sociology of science, as well as explanation and realism, but feminist approaches are also discussed and things are wrapped up with a consideration of future directions.

Hacking, I., *Representing and Intervening*, Cambridge: Cambridge University Press, 1983.

This is part accessible introductory text, part defence of the shift in focus from theory to experiment, and includes numerous interesting (if sometimes rather sketchily presented) examples from scientific practice. Although the discussion of the state of play in the debate over scientific realism is now out of date, it is here that one finds the first detailed articulation of 'entity realism', with its famous slogan 'If you can spray them, then they're real.'

Kitcher, P., *The Advancement of Science: Science without Legend, Objectivity without Illusions*, Oxford: Oxford University Press, 1993.

This book offers a robust defence of the notions of objectivity and progress in science. Kitcher argues that many criticisms of these notions have focused on the idealizing features of science and he offers an account in terms of 'real world' decision making, complete with bias, the impact of social factors etc. Scientific progress is described in terms of a consensual practice which deploys various explanatory resources, and advances via accumulating truths. It is quite challenging but also clearly written and draws on a number of examples from the history of science, including Darwinian evolution.

Klee, R., *Introduction to the Philosophy of Science: Cutting Nature at its Seams*, Oxford: Oxford University Press, 1997.

Yet another useful and accessible introduction that adopts a broadly historical approach, beginning with the positivist view of science and then proceeding through falsificationism, the Kuhnian view and subsequent sociological and feminist accounts. It also covers the problem of underdetermination, realism in general and explanation, although once again it is a little out of date. However, in contrast with many books in this area that tend to draw only on examples from physics, it has some nice examples from immunology.

Klee, R., *Readings in the Philosophy of Science: Cutting Nature at its Seams*, Oxford: Oxford University Press, 1998.

This presents a selection of classic papers covering a wide range of topics, including the theoretical/observational distinction, holism and underdetermination, realism–anti-realism, the (inevitable) Kuhnian model of science and sociological approaches and feminism. Although now somewhat out of date, it still contains useful material.

Kuhn, T. S., *The Structure of Scientific Revolutions*, 2nd edn, Chicago: University of Chicago Press, 1970/62.

This is one of the most influential books in the field, with everyone from American Presidents to comedians using the word 'paradigm', as introduced here. It is written in an accessible, non-technical manner, with lots of examples from the history of science and is generally credited with helping to bring the philosophy of science back into touch with the history of science.

Kuipers, T. (ed.), *General Philosophy of Science: Focal Issues* (Handbook of the Philosophy of Science), Amsterdam: Elsevier, 2007.

Here we have another collection of essays that covers a selection of 'focal issues', ranging from theories of explanation to the role of experiments in the natural and social sciences. This is not exactly introductory material, but the essays still offer broad and useful overviews of some of the central topics in philosophy of science.

Ladyman, J., *Understanding Philosophy of Science*, London: Routledge, 2002.

This is an accessible and engaging introduction that covers all the main topics, beginning with induction and taking the reader through falsificationism, Kuhn's paradigm-based approach (see above!), the realism debate and explanation. The author has made major contributions to the realism debate himself so it should come as no surprise that the discussions of that, as well as constructive empiricism, are particularly good.

Machamer, P. and Silberstein, M., *The Blackwell Guide to the Philosophy of Science,* Oxford: Oxford University Press, 2002.

After an introductory chapter on the history of the subject and another on the 'classic' debates and future prospects, this volume presents a series of specially written essays by leading philosophers of science on such standard issues as explanation, reduction, observation etc., as well as on more specialized topics such as evolution and the philosophy of space-time. It concludes with a chapter on feminist approaches to the philosophy of science.

Massimi M. et al. (eds), *Philosophy and the Sciences for Everyone*, London: Routledge, 2014.

The complementary text to the University of Edinburgh's on-line course that examines the intersection of science and philosophy, this volume begins with a short but engaging philosophical response to the question 'What is this thing called science?' The rest is divided into two parts: the nature of the universe, including dark energy and multiverse cosmology and consciousness, covering intelligent machines and embodied cognition. It thus offers an original perspective on the philosophy of science via debates between philosophers and scientists themselves.

Okasha, S., *Philosophy of Science: A Very Short Introduction,* Oxford: Oxford University Press, 2002.

As it says on the tin, this is a short introduction to the major topics in the field. It begins with a discussion of the distinction between science and pseudo-science, before tackling induction and probability, explanation, the realism debate, Kuhn and sociological approaches, among other topics. The author is one of the world's leading experts in the philosophy of biology, so the discussion of that topic is particularly good and, unusually for an introductory text, philosophical issues in linguistics are also covered. The book ends with consideration of the relationship between science and religion and the so-called 'science wars' between 'rationalist' and sociological approaches to science.

Psillos, S., *Causation and Explanation,* Acumen, 2002.

An accessible and helpful introduction to the major philosophical debates over causation and explanation, this also covers issues to do with the nature of scientific laws. It is historically sensitive and draws upon the relevant metaphysics, displaying an impressive command of the literature.

Psillos, S., *Scientific Realism: How Science Tracks Truth,* London: Routledge, 1999.

This is the classic defence of scientific realism against both the Pessimistic Meta-Induction and the Underdetermination of Theories. Although it is by no means an introductory text the book is clearly written and includes some useful case studies such as the caloric theory of heat and Maxwell's theory of electromagnetism.

Rosenberg, A., *Philosophy of Science: A Contemporary Introduction*, London: Routledge, 2000.

Rosenberg argues that the problems of philosophy of science are among the most fundamental problems of philosophy in general and offers a detailed discussion of explanation, causation and laws. He also covers the structure of scientific theories, the problem of theoretical terms and the nature of theory testing before ending with a consideration of relativism. Useful lists of further reading are included, together with study questions at the end of each chapter.

Schurz, G., *Philosophy of Science: A Unified Approach*, London: Routledge, 2013.

This is a dense and somewhat challenging text that nevertheless contains a rich set of resources. Beginning with an overview of the historical development and general aims of the philosophy of science itself, the core of the book is concerned with establishing a basis for a unified approach to science, taking us through such topics as laws and empirical testing, the empirical evaluation of theories more generally, including issues of realism and empiricism, before concluding with an extensive and technically detailed account of causation and explanation. Chapters are divided into core and 'complementary and advanced' topics and there are some useful case studies as well as lists of related literature and sets of exercises for students at the end of every chapter.

The following are collections of papers, some classic, some contemporary, but which are generally more advanced (again, the descriptions have been adapted from *The Bloomsbury Companion to the Philosophy of Science*):

Balashov, Y. and Rosenberg, A., *Philosophy of Science: Contemporary Readings*, London: Routledge, 2002.

This collection presents a range of classic, and in some cases quite challenging, papers (including one by Charles Darwin) on a variety of issues from 'Science and Philosophy', through 'Explanation, Causation and Laws' to 'Realism' and 'Science in Context'.

Boyd, R., Gasper, P. and Trout, J. D. (eds), *The Philosophy of Science*, Cambridge, MA: The MIT Press, 1991.

This is a chunky volume that contains more than forty original and classic readings in the philosophy of science. Beginning with essays on positivism and proceeding through feminism and science, it also includes sections on the philosophy of particular sciences, such as physics, biology and psychology. With useful introductory essays that help to orient students this remains a handy resource, even if it is now somewhat out of date when it comes to contemporary discussions.

Brown, J. R. (ed.), *Philosophy of Science: The Key Thinkers*, London: Bloomsbury, 2012.

I'm not such a big fan of approaching the philosophy of science via so-called 'key thinkers' but this is interesting as the commentators on the chosen 'key thinkers' might well be described as key thinkers themselves. So, we have Stadler discussing the positivists, Carrier on Kuhn, Lakatos and Feyerabend, Psillos comparing Putnam with the arch-anti-realist, van Fraassen, and Kourany examining the feminist critiques of Harding and Longino. With a useful facing-the-future afterword by Brown, the core themes of the field still emerge strongly and this offers an interesting way of accessing them.

Cover, J. A., Curd, M. and Pincock, C., *Philosophy of Science: The Central Issues*, 2nd edn, New York: W. W. Norton, 2012.

Another brick of a book, this one presents a selection of classic and more recent papers and is divided into nine sections, each book-ended by an introduction and a detailed commentary covering the main issues and core arguments of that topic. The sections are 'Science and Pseudo-Science', which includes essays on an early US court case on creation science; 'Rationality, Objectivity and Values in Science', which includes a discussion of gender and the biological sciences; 'The Duhem-Quine Thesis and Underdetermination', bringing together two issues I have discussed under 'Justification' and 'Realism'; 'Induction, Prediction and Evidence', with a useful initial essay on induction; 'Confirmation and Evidence: Bayesian Approaches'; 'Models of Explanation', which covers both older and more recent approaches; 'Laws of Nature'; 'Intertheoretic Reduction'; and 'Empiricism and Scientific Realism', which includes papers on the Pessimistic Meta-Induction as well as critical analyses of constructive empiricism and 'entity realism'. In my honest opinion, this is one of the best such anthologies on the market.

French, S. and Saatsi, J., *The Bloomsbury Companion to the Philosophy of Science*, London: Bloomsbury, 2011/14.

Not surprisingly, I think this is ace! The core of the book features specially commissioned essays by leading philosophers of science on topics from its relationship with epistemology, metaphysics and the history of science, to realism, reduction, explanation, causation, confirmation and the nature of evidence. It also includes papers on the philosophy of particular sciences, such as biology, chemistry, physics and mathematics, but also, unusually for this sort of volume, economics and neuroscience. There is a useful, if brief, chronology of the history of the subject, as well as a concluding essay on future possible developments and a list of research resources, including blogs as well as books. It's sure to become a classic!!

Hitchcock, C. (ed.), *Contemporary Debates in the Philosophy of Science*, Oxford: Wiley-Blackwell, 2004.

This is a somewhat unusual collection, in terms of both form and content. It presents eight disparate issues in philosophy of science, organized in a debate format, with two papers defending explicitly opposite positions on each issue. This is a nice way of bringing the issues alive but although the participants include some of the very best philosophers of science around, it has to be admitted that some of the interactions are less engaging than others. The topics covered are also a little idiosyncratic perhaps (e.g. 'Do Thought Experiments Transcend Empiricism?') but there is some very accessible philosophy here, and it is worth dipping into.

Lange, M., *Philosophy of Science: An Anthology*, Oxford: Wiley-Blackwell, 2006.

This is another sweeping selection of classic papers, beginning with empiricism and including essays on confirmation, theory choice, realism, causation and explanation, as well as a number of others. It also includes a section on the metaphysical implications of modern physics. Not surprisingly, given the choice of essays, it overlaps quite considerably with the other leading anthologies but with 38 articles in total this is one of the most comprehensive.

Machamer, P. and Silberstein, M., *The Blackwell Guide to the Philosophy of Science*, Oxford: Oxford University Press, 2002.

After an introductory chapter on the history of the subject and another on the 'classic' debates and future prospects, this volume presents a series of specially written essays by leading philosophers of science on such standard issues as explanation, reduction, observation etc., as well as on more specialized topics such as evolution and the philosophy of space-time. It concludes with a chapter on feminist approaches to the philosophy of science.

Newton-Smith, W. H. (ed.), *A Companion to the Philosophy of Science*, Oxford: Blackwell, 2001.

With a list of topics from 'Axiomatization' to 'Whewell' and a range of contributors from Worrall to Achinstein this offers an encyclopaedic series of snapshots of issues in the field written by leading researchers. Although some of the issues covered are not as central as they were once thought to be, this is still a useful guide for the beginner and a handy reference work for the more advanced student.

Papineau, D. (ed.), *The Philosophy of Science*, Oxford: Oxford University Press, 1996.

This is mainly focused on the realism–anti-realism debate. As in other cases, many of these papers are no longer at the cutting edge, and some are quite challenging for the undergraduate student. Nevertheless, it contains some classic pieces, such as Worrall's original paper on structural realism, Laudan's presentation of the pessimistic meta-induction and van Fraassen on saving the phenomena. Papineau also provides a helpful orientation to the debate in the introduction.

Psillos, S. and Curd, M., *The Routledge Companion to Philosophy of Science,* London: Routledge, 2008/13.

This contains over sixty specially commissioned essays on a range of topics and issues in the field. Part 1 covers the 'Historical and Philosophical Context', with essays on metaphysics, the role of logic in the philosophy of science and its relationship with the history of philosophy. Part 2 is simply entitled 'Debates' and includes essays on computer simulation, the ethics of science, relativism, feminist approaches and theory change, plus many more(!). Part 3, 'Concepts', covers a huge range from causation, idealization, models, representation, reduction and symmetry, among others. The final part examines the philosophy of the 'individual' sciences, such as biology, physics, cognitive science, economics and the social sciences. Recently updated, this is a rich and up-to-date reference work that provides a useful entry point to the most pressing debates and issues in the subject.

Suggested readings

Chapter 2: Discovery

Nickles, T., 'Scientific Discovery', in Curd, M. and Psillos, S., *The Routledge Companion to the Philosophy of Science*, London: Routledge, 2008, pp. 442–51.

Nickles was one of the first philosophers of science to reject Karl Popper's admonition that discovery was not a fit topic for philosophical analysis and argued that methods of discovery could be set alongside forms of justification of theories. This 'companion' piece presents an overview of the issues and offers his own take on them.

Schickore, J., 'Scientific Discovery', *The Stanford Encyclopedia of Philosophy* (Spring 2014 Edition), Edward N. Zalta (ed.). Available from http://plato.stanford.edu/archives/spr2014/entries/scientific-discovery/

This is a comprehensive and generally accessible survey of philosophical views of scientific discovery that covers the 'Eureka' moment and heuristics-based accounts, as well as the distinction between contexts of discovery and justification. It also examines discoverability as a mode of justification and concludes with discussion of psychological research on creativity and mental models.

Waller, J., *Fabulous Science: Fact and Fiction in the History of Scientific Discovery*, Oxford: Oxford University Press, 2004.

An accessible and historically rich account of several important scientific discoveries that draws on examples from the history of medicine and biology, as well as Eddington's observation of light bending round the sun. Clearly written by a historian of medicine, this book bursts a number of popular bubbles.

Chapter 3: Heuristics

http://en.wikipedia.org/wiki/Heuristics_in_judgment_and_decision-making

A useful survey article that covers all the main heuristic devices and associated biases, from representativeness to the base rate and conjunction fallacies.

Kahneman, D., *Thinking, Fast and Slow*, London: Allen Lane, 2011.

This is the Nobel prize winner's best-selling analysis of decision making in terms of 'fast' and 'slow' thinking, the former being intuitive but biased and the latter slow but logical. Kahnemann argues for the role of both inappropriately framing risks and making decisions across a range of issues.

Chapter 4: Explanation

De Regt, H. W., 'Explanation', in French, S. and Saatsi, J. *The Bloomsbury Companion to the Philosophy of Science*, London: Bloomsbury, 2011/14, pp. 157–78.

This is an accessible and engaging survey of all the major accounts of explanation, from the D-N and causal views, to functional approaches. It also looks at explanation in the human sciences and the pragmatics of explanation, and concludes with a suggestive projection of future research.

Mayes, G. R., 'Theories of Explanation', *Internet Encyclopaedia of Philosophy*. Available from http://www.iep.utm.edu/explanat/

Although there is a heavy emphasis on the D-N account, this is a useful encyclopaedia article that covers a range of contemporary approaches and includes a section on explanation in cognitive science.

Woodward, J., 'Scientific Explanation', *The Stanford Encyclopedia of Philosophy* (Winter 2014 Edition), Edward N. Zalta (ed.). Available from http://plato.stanford.edu/archives/win2014/entries/scientific-explanation/

An outstanding survey by someone who has made major contributions to the field, which again covers the D-N and causal accounts, as well as pragmatic and unificationist approaches. It concludes with a discussion of how explanation meshes with other goals of inquiry and whether a single model of explanation is desirable or even possible.

Chapter 5: Justification

Hawthorne, J., 'Inductive Logic', *The Stanford Encyclopedia of Philosophy* (Winter 2014 Edition), Edward N. Zalta (ed.). Available from http://plato.stanford.edu/archives/win2014/entries/logic-inductive/

Although this article covers a lot of technical material, the first three sections are reasonably introductory, as Hawthorne himself notes, in setting out the differences between deductive and inductive logic before taking the reader by the hand and leaping off the cliff into the seas of Bayesian estimation and 'likelihood ratio convergence' (!).

Norton, J., 'A Little Survey on Induction', in P. Achinstein (ed.), *Scientific Evidence: Philosophical Theories and Applications*, Baltimore: John Hopkins University Press, 2005, pp. 9–34 (also available here: http://www.pitt.edu/~jdnorton/papers/Little_Survey_Final.pdf)

Norton's 'little survey' is comprehensive but also, like many essays on this topic, quite challenging. His 'schtick' is that unlike the case of deductive logic, we may just have to accept that there exist many inductive schemas, each appropriate to a specific, local context of enquiry.

Popper, K. R., 'Science: Conjectures and Refutations', a 1953 lecture which forms the first chapter of his book *Conjectures and Refutations*, London: Routledge, 1963/2002, but which can also be found separately on the web.

Sometimes there's no better option than to go straight to the source! In this essay, originally presented in Cambridge, Popper begins with some autobiographical reflections on his training in psychology (including the infamous example of the drowning child) before considering the problem of how we should distinguish science from pseudo-science and proposing his criterion of falsifiability as a solution. After dismissing inductive reasoning he concludes by suggesting that science proceeds via a process of conjectures and refutations.

Chapter 6: Observation

Bogen, J., 'Theory and Observation in Science', *The Stanford Encyclopedia of Philosophy* (Summer 2014 Edition), Edward N. Zalta (ed.). Available from http://plato.stanford.edu/archives/sum2014/entries/science-theory-observation/

Beginning with the positivists' emphasis on the objectivity and accessibility of observation reports – in part as a response to the Nazis' anti-semitism – Bogen covers a wide range of associated topics, including not only the way observation relates to theory and how it may be said to be 'theory laden' but also whether observation is based on perception only. Some nice examples from science are used to illuminate crucial points, including magnetic resonance imaging of the brain and data on orgasms in female macaques. The article concludes with a discussion of the distinction between data and phenomena that draws on the example of electrical activity in nerve fibres.

Kukla, A., 'Observation', in Curd, M. and Psillos, S., *The Routledge Companion to the Philosophy of Science*, London: Routledge, 2008, pp. 396–404.

An excellent survey piece that examines, in particular, different attempts to delineate the observational features of scientific practice and relates these to the debate between realists and anti-realists. It also gives a useful set of further readings.

Chapter 7: Experiment

Franklin, A., 'Experiment in Physics', *The Stanford Encyclopedia of Philosophy* (Winter 2012 Edition), Edward N. Zalta (ed.). Available from http://plato.stanford.edu/archives/win2012/entries/physics-experiment/

Written by one of the first people to really take a serious look at how experiments work, this offers some general conclusions about the role of experiment, despite focusing mainly on examples from physics. In particular, it begins with Hacking's interventionist account (see below) but then points out that it is incomplete and continues by elaborating the various roles of experiment before concluding with sections on computer simulations as experiments and experimentation in biology (see Weber's entry, below).

Hacking, I., *Representing and Intervening*, Cambridge: Cambridge University Press, 1983.

This has become a classic in the philosophy of science, not least because it not only introduces and defends a provocative idea – that experimentation in science involves not just observing but *intervening* in the world – but it is also very clearly written and accessible. Some of the discussion – e.g. of realism – is now a touch out of date but the examples are useful and the core idea remains a powerful one.

Weber, M., 'Experiment in Biology', *The Stanford Encyclopedia of Philosophy* (Winter 2014 Edition), Edward N. Zalta (ed.). Available from http://plato.stanford.edu/archives/win2014/entries/biology-experiment/

As Weber notes, a large part of the philosophy of biology has traditionally been taken up with looking at biological theories and models but, due to the impact of historical studies, philosophers of biology have shifted more attention to the nature and role of experiment. Beginning with the issue of how causal relationships can be inferred from biological data, Weber discusses theory testing and confirmation in biology before moving on to the role of 'model' organisms and the reliability of biological data.

Chapter 8: Realism

Chakravartty, A., 'Scientific Realism', *The Stanford Encyclopedia of Philosophy* (Spring 2014 Edition), Edward N. Zalta (ed.). Available from http://plato.stanford.edu/archives/spr2014/entries/scientific-realism/>

This offers a comprehensive and detailed survey of the realism/ anti-realism debate, beginning with what it is to be a scientific realist and what are the arguments in favour, before moving on to the arguments against and the various forms of anti-realism. What's particularly useful about the latter sections is that they include not only empiricist forms of anti-realism but also feminist and pragmatic accounts.

Psillos, S., *Scientific Realism: How Science Tracks Truth*, London: Routledge, 1999.

The classic defence of 'standard' scientific realism. I would skip the first four chapters as these are mostly historical but from chapter five onwards we get a careful and well-argued defence of realism against the pessimistic meta-induction and underdetermination arguments that draws on examples from the history of science such as the caloric theory of heat. It also includes critical discussions of constructive empiricism and structural realism and touches on other philosophical issues such as the reference of theoretical terms.

Wylie, A., 'Arguments for Scientific Realism: The Ascending Spiral', *American Philosophical Quarterly* 23 (1986): 287–98.

Although this is also out of date in terms of the various forms of realism and anti-realism that have been developed over the past thirty years, Wylie's paper is still a very nice, accessible piece that helps explain why the debate has continued for so long and why the participants continue to spiral around one another. It can usefully be read in conjunction with Chakravartty's piece, above.

Chapter 9: Anti-realism

Chakravartty, Anjan, 'Scientific Realism', *The Stanford Encyclopedia of Philosophy* (Spring 2014 Edition), Edward N. Zalta (ed.). Available from http://plato.stanford.edu/archives/spr2014/entries/scientific-realism/

See the suggested readings for Chapter 8 but let me just add: Chakravartty also includes discussion of the likes of Kuhn and the social constructivists and ends with a consideration of the question whether the dispute between realism and anti-realism is even resolvable (also, see Wylie's paper given in the readings for Chapter 8).

Monton, B. and Mohler, C., 'Constructive Empiricism', *The Stanford Encyclopedia of Philosophy* (Spring 2014 Edition), Edward N. Zalta (ed.). Available from http://plato.stanford.edu/archives/spr2014/entries/constructive-empiricism/

As the authors of this article note, constructive empiricism is widely viewed as having rehabilitated scientific anti-realism, but the view has evolved and developed over the years in response to criticisms and further sustained reflection on the nature of empiricism regarding science. This encyclopaedia essay offers a useful account of those developments, as well as the arguments for and against, both good and bad. It finishes with a consideration of what such an anti-realist about scientific entities might say about mathematical ones.

Chapter 10: Independence

Bloor, D., 'Relativism and the Sociology of Scientific Knowledge', in Hales, S. D. (ed.), *A Companion to Relativism*, Oxford: Wiley-Blackwell, 2011, pp. 433–55.

Beginning with Kuhn's philosophy of science, this survey essay from one of the founders of the so-called 'strong programme' poses the question, Why should we insist that scientific reasoning must be disentangled from the social 'influences' that supposedly impact upon it? It then helpfully distinguishes various senses of 'relativism' before setting out the arguments for and against in a clear but engaged manner and ends with a plea to at least be charitable in critically assessing relativists' claims.

Longino, H., 'The Social Dimensions of Scientific Knowledge', *The Stanford Encyclopedia of Philosophy* (Spring 2013 Edition), Edward N. Zalta (ed.). Available from http://plato.stanford.edu/archives/spr2013/entries/scientific-knowledge-social/

Longino begins with some historical background before moving into issues to do with the emergence of 'big' science and the loss of trust as motivating factors in the recent development of social and cultural studies of science. The article presents an extensive bibliography and concludes with a survey of more recent work and an indication of future directions of research.

Chapter 11: Gender bias

Anderson, E., 'Feminist Epistemology and Philosophy of Science', *The Stanford Encyclopedia of Philosophy* (Fall 2012 Edition), Edward N. Zalta (ed.). Available from http://plato.stanford.edu/archives/fall2012/entries/feminism-epistemology/

Here Anderson sets feminist philosophy of science in the context of a feminist approach to philosophy more generally, beginning with the idea of knowers as 'situated' in a particular (e.g. gendered) context. She then takes the reader through feminist standpoint theory, empiricism, post-modernism and science theory, before discussing feminist critiques of objectivity and authority regarding knowledge. Her essay concludes with a look at the recent convergence between these various approaches as well as external criticisms and feminist responses.

Crasnow, S., 'Feminist Philosophy of Science: Values and Objectivity', *Philosophy Compass* 8 (2013): 413–23.

Crasnow begins by noting that although feminist philosophers of science have been in the forefront of critiques of the supposed value-free nature of science, they also have an interest in retaining some measure of objectivity. Setting out various accounts that attempt to meet this challenge, Crasnow covers feminist empiricism and standpoint theory, before concluding with a useful summary and suggestions for future feminist research.

Wylie, A., Potter, E. and Bauchspies, W. K., 'Feminist Perspectives on Science', *The Stanford Encyclopedia of Philosophy* (Fall 2012 Edition), Edward N. Zalta (ed.). Available from http://plato.stanford. edu/archives/fall2012/entries/feminist-science/

As the authors note, feminist perspectives on science reflect a broad spectrum of attitudes towards, and appraisals of, science. In an extensive survey article, they map out the differences between these perspectives along various dimensions and compare the appropriation by feminists of scientific methodology to cover areas previously neglected by (male) scientists with deeper feminist critiques of that methodology. They also consider the philosophical implications of such critiques for issues such as objectivity and confirmation and conclude with a useful summary and a stirring dismissal of the view that introducing values into science corrupts its reliability.

Index